WHICH ONES ARE SCIENTIFIC?

Evolution, Creation, I.D. or Hybrids, through Figures & Tables

By George Grebens

T0130146

 www.trafford.com

North America & international
toll-free: 1 888 232 4444 (USA & Canada)
phone: 250 383 6864 ✦ fax: 250 383 6804 ✦ email: info@trafford.com

The United Kingdom & Europe
phone: +44 (0)1865 487 395 ✦ local rate: 0845 230 9601
facsimile: +44 (0)1865 481 507 ✦ email: info.uk@trafford.com

10 9 8 7 6 5 4 3 2 1

WHICH ONES ARE SCIENTIFIC?

Evolution, Creation, I.D. or Hybrids, through Figures & Tables

By George Grebens (PhD)

<u>Dedication</u>

To my wife

and

To the countless saints who have dedicated their lives and offered the ultimate sacrifice during the 20th Century (500 million+).

WHICH ONES ARE SCIENTIFIC?

Evolution, Creation, I.D. or Hybrids, through Figures & Tables

By George Grebens

Table of Contents

WHICH ONES ARE SCIENTIFIC?

Evolution, Creation, I.D. or Hybrids, through Figures & Tables

By George Grebens

Figures and Tables

Tables

PREFACE

This book is a *summary* of my original 330+-page book, with over 288 footnotes and 3 appendices. That research entitled '*Evolution, Creation, And Intelligent Design: Which ones are scientific?*' (*ECID*), 2009, contains nine of the 35 Figures and Tables in the standard format, while the remaining 26 appear in text format due to costs of publication. This second book, entitled '*Which Ones are Scientific? Evolution, Creation, I.D. or Hybrids – Figures and Tables* (*WOAS*), meets the following objectives:

a) Include and describe all of the 35 Figures and Tables in their original format.
b) *Summarize* the original research work (*ECID*) on 178 pages. This summary (*WOAS*) becomes a reference manual with some essential, and some new 80 footnotes and 2 appendices.
c) The 35 Figures, Tables, and lists as well as supporting text form the content of the *Scientific Score Card*. This *Score Card* allows the reader to compare and evaluate the scientific merit of what the debaters offer.
d) Provide additional clarifying examples, methods, points and end notes.

The reader can easily recognize the four debating groups – a) '*Neo-Evolution*', b) '*Young Earth Creation*', c) '*Intelligent Design*' and the d) *neo-Theistic Evolution* or '*Hybrid Uniformitarians*' (i.e., '*Progressive Creation*' or '*Old Age Creation*').

In addition to these 4 debaters, there are, hundreds of thousands of additional participants, whose roles on these social and civilization issues range from administrative proponents, financial sponsors and promoters, courts judges, two types of scientists (professional and ideological), lab specialists, university professors, teachers, students, curriculum developers, religious and political organizations, the media, as well as, activists and website debunkers. All of these directly or indirectly participate in the movements that the debaters represent. Inevitably, the debate also involves millions around the world – the wider audience that examines the content of the debates, evaluates, asks questions, passes judgment, and expresses opinions and preferences.

Both books - *ECID* and *WOAS,* do not only reflect mainstream themes, strategies and arguments, but also go beyond by examining historical trends that help identify the source, content, application and management of the original knowledgebase, science, the scientific method. This historical trend also includes the framework of myths, philosophies, religions, and ideologies. All of these, coupled with economics, have become the vehicle that has navigated the scientific knowledgebase across civilizations. Such a historical perspective provides a better grasp of the original definitions that led to our modern world, which increasingly today confines itself to the narrow uniformitarian scope.

To facilitate this complex subject, the reader will not only have the opportunity to use up to 35 Figures and Tables and additional lists as a comparative *Scientific Scorecard,* but this scorecard will also allow the reader to identify objective criteria that lie beneath the content of the debates. For example, the reader would wish to determine which of the debating groups come closest to an 'objective and unfettered science and the scientific method' (OUSM)? What are the five qualitative change levels? How are these variables (science, scientific method, ideology, religion, etc.), managed? What are the five basic models of thinking? What are the 12 limitations of science and of the scientific method? What are the 22 most common violations of the scientific method? And many other topics.

In addition, to gain an objective perspective, the two books apply techniques used in the management of multi-billion dollar engineering projects. The two books have simplified this management practice in the form of a *three dimensional management model* (3DMM). This 3DMM foundational data, information and knowledgebase structure exists not only in science, the scientific method, but also in the management of variables in philosophy, ideology, religion - which function in the background during the debates. The debaters unconsciously integrate their positions within these structures.

The *3D Management Model* helps accomplish three goals: 1) manage information through three infrastructures (planning, operations and style) and 27 management plans. This network converts information into a structured knowledgebase. 2) With this, the users' objective is to formalize a body of accurate knowledge (science) and contribute to 'conditional certainty.' 3) Whenever knowledge experts identify or predict 'uncertainty' – it is then the *8-point Decision-Making and Problem-Solving* (DMPS) *method* (scientific method) that helps resolve 'uncertainty.' In other words, the aim of DMPS (scientific method) is to convert 'uncertainty' to 'conditional certainty' – a more accurate knowledgebase – science, i.e., 3DMM (3-Dimensional Management Model).

Both books provide a historical perspective on how science had developed from Sumerian times to our own. Civilizations had to maintain specific favorable conditions to allow the development of science, the scientific method, and technology. In addition, such a favorable condition would allow for the transmission of this body of accurate knowledge from one civilization to the next. In contrast, it will then be easy to identify those historical trends that had brought about unfavorable conditions, and closed or distorted opportunities for understanding what science is, and misapplied the scientific method. Such conditions resulted in reduced naïve realism and brought about the dark ages.

INTRODUCTION
Science from the Dawn of Time to Our Own

Unbelievably, science and the scientific method had been evident since the dawn of time. Perhaps the earliest scientific instrument that we still have with us had been the zodiac. It is a 6000-year-old application that anyone, anywhere on the surface of the Earth, in both hemispheres, can see and use nightly. Navigators have used these grouped constellations to plot routes on the high seas, in uncharted territories, on deserts trails and in mountainous regions. Astronomers used these to compute sophisticated stellar events, calendars and plotted accurate scientific observations and made forecasts.

Science is also evident in the 4000+ year old maps. Discovered in the Turkish Library (Istanbul/ Constantinople) these maps reveal ancient land contours of all of the Earth's continents - including the Antarctic continent. There are two sets of these ancient maps – the first detailed all continental contours when sea levels were low and the Antarctic continent was de-glaciated. Another set of maps described the same continents when in circa 1500 B.C. sea levels rose 500 feet at the equator, and 100 feet on latitudes further north and south[i].

We have recognized pyramids in the Middle East and Latin America, the numerous megalithic structures, and Greek architecture. These civilizations developed methods for resolving uncertainty, reduce risk and take advantage of opportunities. Both economics and the 'myth' had provided the vehicle for the development and transmission of the scientific knowledgebase. From the beginning, we see geometry in astronomy, geography, maps, navigation and military engineering, architecture. Evident also are detailed accounting systems in trade, tax management, legal documentation, rules, agriculture, animal breeding, and even inventions such as Greek fire, electric batteries and the numerous mechanisms that predate the robotic systems.

European astronomy and mathematics, can trace their scientific roots among Sumerian cuneiforms. Here, recorded are a numeric system based on sexagesimal (base 60) place value, a circle divided into 360 degrees of 60 minutes each. The Greek Hipparchus' borrowed and wrote about this in his 'Almagest', 'Planetary Hypotheses', 'Tetrabiblos' and 'Canobic Inscription'. There is evidence that Babylon I & II and Egypt inherited catalogs of stars with constellations, which include lunar and planetary predicted movements about the horizon. Similarly, water clocks, gnomon, shadows, intercalations had been used in those early times. Mathematics included arithmetic, algebra, geometry, and trigonometry. From Babylon, Egypt to Persia, and through the exploits of Alexander the Great in these geographic areas, Greeks inherited this knowledge. This included the heliocentric model of planetary motions, which we find identified in Seleucus of Seleucia's work (190 B.C.). Here we find the Metonic cycle – a lunisolar calendar based on 19 solar years (235 lunar months - the synodic month).

Similarly, Hipparchus and later Ptolemy maintained a multi-year complete list of eclipse observations that helped track dates from 652 BC to 130 AD.

Greek astronomy had been a branch of mathematics – geometric models that led to the creation of a two-sphere model that represented stellar and planetary phenomena. This Hellenistic knowledge also entered India (Greek-Bactrian city of Ai-Khanoum – 3rd century B.C.) and developed in the 6th century in the 'Romaka Siddhanta' – Doctrine of the Romans, and others.

In the 3rd century B.C. the major center of research and learning – the library of Alexandria, Egypt - established under Ptolemy II, had existed to serve several empires, until this center moved to Byzantium and Rome. For example, we have three volumes written by the mathematician and physicist Hero (Heron or Heronis) (10 – 70 AD) of the Alexandrian Library, who experimented with, devised and documented technological achievements in automation, programmable machines (his automated theater) that included not only cogs, wheels and strings but also hydraulic systems, pneumatics and robotics[ii]. He first documented steam-powered devices, the aeolipile (a rocket-like device driven by steam), the first vending machines, optics that are examined in his *'Principles of the Shortest Path of Light,'* and wind-wheel[iii] [iv] [v]. In Alexandria he represented the Hellenistic scientific tradition.

Later, this knowledgebase became foundational for Islamic astronomy and mathematics. For example in 1030 A.D. Al-Biruni in his 'Indica' discussed Indian astronomical theories of Aryabhata, Brahmagupta and Varahamihira – specifically about Earth's rotation on its axis, within two possible views: geo or heliocentric ones. Abu Said Al-Sijzi narrowed the options by proposing that the Earth moved around the Sun.

 The Byzantine Empire maintained a scientific knowledgebase that was brought to the West between 1205 (Fourth Crusade) and 1454 (Ottoman takeover of Constantinople). Western Medieval Christian Europe also received some Greek and Hindu knowledge through translations that originated in Moorish Spain. In Western Europe we can site the works of Cardinal Nicholas of Cusa ('Learned Ignorance' –'De docta ignorantia') 15th century, and Nicolaus Copernicus in his 'De revolutionibus' (1543) that cite theories of Albategni, Arzachel and Averroes.

During the recent 500 years, there has been an increasing tendency to re-interpret and rewrite European history. What remains true, however, is that the Christian classical foundations have helped maintain a civilization that was favorable for the development of science and the scientific method. The Byzantine Empire (Constantinople) sustained an effective administration, military strength and economic trend in the Eastern Mediterranean Roman Empire. When in the 5th and 6th centuries the Western Roman Empire underwent challenging conditions, it is the Christian Church of Rome and the Christian Celtic culture that helped maintain the civilizing anchors. From this,

Charlemagne's empire and the Christian Celtic culture that offered a European network with a better civilizing option. These Christian civilizations continued to offer an organizing and cultural option throughout Europe for two millennia.

This Christian unifying option provided favorable conditions for identifying and establishing discoverable laws within a divinely ordained creation. Universal lawfulness contributed to the discovery of accurate, certain knowledge that led to technological progress. Furthermore, this universal lawfulness had led to the continuous self-perfection of man and society.

Upon closer examination, history has shown that accurate knowledge (science) relied on the view of some kind of lawful universe and of society. This has become the common thread wherever civilizations rose. Favorable conditions for the development of accurate knowledge necessitated a worldview where the heavens, earth and economics reflected laws through which it had been possible to make accurate predictions. It is here that data, information, standards, codes had become key to accurate observation. These observations helped define, compare alternatives. This knowledgebase reflects codes, documentation, standardization, and tests results for quality and reliability. Finally, such a civilization manages its knowledgebase with management style: meaningful attitude, motivation, culture and ethics.

Christian civilization reflected these qualities. It continuously tracked and resolved uncertainty. It contributed to creative activities such as **engineering projects** (planned and scheduled resources to meet objectives); **automation** (standardized machine and clock processes that yielded high quality products) and **manufacturing** (conversion of prime resources to final quality products). Here the Biblical Creator God created the Heavens and the Earth established reality upon geometric natural law. This law is discoverable since it helps identify and resolve uncertainty. In other words, this law contains the necessary *management* features that help identify and convert uncertainty to conditional certainty. This geometric natural law provided the strategic means that helped identify workable, measurable and reproducible solutions. These features then helped open the road to further layers of qualitative discoveries of measureable and recognizable knowledge. These two integral features helped establish manageable infrastructures, standards and means to calibrate, thus contributing to a third level of discoverable infrastructural laws. In this three-layered array, 3DMM identifies ethics, culture, and attitude among others as chief components of the knowledgebase and means to resolving uncertainty. These, in turn, affect society's mission, economic, health that sustain and lead it to self-perfection.

European historical, economic and demographic challenges (e.g., ravages of the black plague in the 14th century west, and 6th century Byzantium) did not dampen an underlying *optimism* that was attached to discoveries. Christian civilization had an integral hope that it would overcome environmental, state and individual limitations and shortcomings. Monasteries and then the medieval universities contributed to the cultural and scientific development in literature, law, architecture, technology, navigation. This in turn led to the earlier 13th century industrial revolution based on wood and stone technology - windmills, watermills, manufacturing, textiles, transportation, clocks, irrigation, astronomical validation, etc. During the 200 years prior to the fall of Constantinople (1453), an exchange of ideas initiated a momentum that spearheaded the Western European Renaissance. This renewal was promoted by the patronage of commercial families, ruling monarchs and its aristocracy, the Pope and the Magisterium (Council of Bishops). Projects included: refinement and recalibration of stellar calculations in conformity with the long known heliocentric view; calendar adjustments contributed to navigation and exploration to China (Marco Polo), the ushering of the printing press and application of gun powder, and the discovery of the new world. Commercial dividends, advanced military technology multiplied and ensured advantages and minimized risks in all of these areas. We can see this in the rediscovery of classical and new architectural designs, building materials (e.g., cement). Here, the Church of Rome (eventually the Vatican) library became a knowledge base of Europe. This Christian millennial timeline has exhibited an optimistic and constructive trend that lasted through the Renaissance and into the twentieth century. Scientists like Leonardo DaVinci, Johann Kepler to Bernard Riemann have spearheaded geometric natural law-based science.

On the other hand, *relativism* drove the unfavorable conditions that helped bring about the dark ages. Historically, relativist worldviews have presented oversimplified reality (naïve realism) that functioned within *rules* rather than *laws* - a relatively prioritized algebraic, axiomatic, reductionist and subjective views. This naïve realism revealed relativism's true weaknesses and objectives. Instead of investing into qualitative infrastructural abundance, the relativists increasingly resign themselves to incremental austerity directives – to do 'more with less' and with the application of 'appropriate technology'. As demonstrated in more detail below, in science, economics and government programs, the relativists promote uncertainty under its many guises. For example, today, when relativists talk about the 'hypothesis' and 'theory' – these never emerge from the first descriptive stage. Such hypotheses and theories never demonstrate the means that lead to conditional certainty. Relativists find that 'evidence' that would lead beyond the initial hypothetical descriptive stages remains 'contaminated', 'inconclusive', and always lacks sufficient 'conditional or quality certainty.' Relativism instead leads to an '*almost* certainty' without experimental proof. Relativists tend to reduce complexity to the lowest common denominator ('reductionism') which is conceived within the mechanical or organic model of thinking.

Reviewing these, unfavorable and non-conducive methods to civilization (see details in Chapter 4.2); we find that all such perceptions and awareness exclude the full range of qualitative and infrastructural organization. Here, relative and reduced reality remains in constant flux (change). The measures for the past and future are simply those of the current systems, processes and rates. Oversimplification ascends above simplification; an alternate interpretation is preferred to qualitative alternatives, and a well-marketed explanation becomes more convincing than an management identified cause, reason and purpose.

Failed civilizations and declining nations with economic meltdowns clearly stand in contrast to developing civilizations that reflect geometric natural law. Characteristically, zero-growth and declining civilizations prioritize subjectivism, decreased respect for human rights, cynicism, institutionalized collectivism, slavery, population reduction schemes, genocide, cannibalism, while court decisions focus on protecting variations of phallic worship. These stand in stark contrast to views that 'man is made in the image of God.'

For the last 500 years, philosophers have used non-traditional methods to attain 'certainty'. These thinkers solely used rationalistic and empiricist methods and tools. In less than 300 years, these initiatives led to the Age of Enlightenment that branched into two opposing views. The first appeared in the form of the American War of Independence, while the second in the French Revolution. The first reflected the principles of the Society of Cincinnatus, the best aspirations of the European Renaissance, geometric natural law, Apostolic Christianity and Alexander Hamilton's *'Report on Manufacturers'* and his nationally coined money project documented in his *'Report on the Establishment of a Mint'*. While the second group led to elevating the Goddess of Reason, use of algebraic, axiomatic rules; promoting perpetual social revolution and colonialism; Adam Smith's *Wealth of Nations* and international banking usury.

A distinction exists in the use of term 'algebra.' This term may refer to a legitimate mathematical discipline, as well as, refer to the algebraic ideological starting premise, which includes axioms, theorems, syllogisms and lies in contrast to the geometric natural law approach. Historically, this ideological algebraic approach has reflected a subjective and relative formula that represented the foundations of failed civilizations and dark ages. In contrast to this, geometric natural law foundations helped create qualitatively continuously improving civilizations with a scientific knowledgebase.

Leading into the 20th century, the ideological algebraic mode removed any semblance of the geometric mode. In the sciences, only a whisper of the geometric school remains.

In the 19th century, we see the emergence of various uniformitarian/evolutionary movements of which Classical Darwinist became dominant towards the latter half of the 19th century. This movement, erected upon the uniformitarian platform, came to reflect up to four axiomatic positions:

1. Start from today's conditions, processes, and rates
2. Through statistical progression, using the simple-to-complex (primitive-to-modern) process, (long ages that allow for statistical progression)
3. Use materialist and reductionist filters – closed system, and exclude open system that are designed to introduce external participating or creative forces
4 Affirm contemporary reality (#1)

By the 1930's Classical Darwinism solidified itself as an *'Evolutionary Synthesis.'* During these early decades, the evolutionary movement made a few tragic social experimental detours through Marxism and Nazism. However, in the 1940s Classical Darwinism in Western Europe and the USA emerges as *Neo Darwinism* or *Neo-Evolution*. By 1983, Neo-Evolution attempts to upgrade itself to *'Methodological Naturalism.'* In the 1980s, and more specifically during the 1980s, the movement revisits its earlier synthesis and reinvents itself in the form of a *'Modern Evolutionary Synthesis.'*

During this same period in the 1980s, from among the Neo-Evolution scientists, while using the new high tech labs some *Neo-Evolution* scientists discovered a myriad of functioning complex information processes, software, and nano-robotics (nanobotics) at sub-chromosome levels. These designs were evident in all living cells (human, animal and plant). These former Neo-Evolution scientists organized themselves as *Intelligent Design (I.D.)* scientists. They recognized that their neo-evolutionary *organic model* could not predict or explain this *Very High Complexity*. I.D. scientists therefore sought an alternate scientific model. The flexible *information-based scientific model* would help them continue legitimate scientific research. Today, however, the *Neo-Evolution* scientists and theorists still promote new editions and versions of the 19th century *organic model*. It is their attempt to salvage the Darwinian reductionist uniformitarian /evolutionary theory.

On the other hand, scientists who in the 19th century did not adhere to the Darwinian view took into consideration the Creation and catastrophic view – the Flood as described in the Christian Biblical legal history. Such scientists interpreted geologic evidence in term of effects of global oceanic tidal fluctuations, temperature inversions, evaporation rates, ocean current velocities (up to 80 km/hr), mega volcanic activity, extensive sedimentation and stratification, rapid fossilization processes; various dating methods, etc. Some 19th century scientists expressed a teleological view (similar to that expressed by the Intelligent Design scientists).

In late 19th and early 20th centuries, some Creation scientists succumbed to the Darwinian 'long age' theory. They incorporated 'long ages' by creating several hybrid uniformitarian explanations such as the -

1) *Day-Age* (i.e., that each day in the Genesis Creation week represents undetermined ages – eventually appeared in three versions)

2) *Gap Theory* (there is an undetermined amount of time between the Biblical Genesis 1:1 and 1:2)

3) *Theistic Evolution* (i.e., several divine mini special creations adjusted evolution where points where fossil gaps were identified in the geologic column)

In the 1950s, the U.S.A government massively invested into modern scientific high tech labs. This resulted in the collection, testing and documentation of voluminous empirical data. Evolutionists had a field day interpreting this data from their uniformitarian perspective. However, beginning with the 1960s Henry M. Morris, who can be viewed as the father of the *modern 'Creation Scientific Model,'* helped provide an alternate interpretation for the massive scientific data. He developed a *Creation Scientific Model* that included *most* of the assumptions maintained by the two millennial Christendom – i.e., the Christian Bible contained accurate and reliable legal historical records of events; reflected natural law, documented three overriding universal singularities that changed all original universal standards. The third singularity – evidence of the global Flood, is corroborated by evidence identified and predicted in geology, paleontology, climatology, archeology, etc.

During the 1970s and early 1980s Creation scientists (www.icr.org - the *Institute of Creation Research*) successfully debated key Neo Evolution scientists at high-level public debates. Eventually, various Neo-Evolution authorities strongly discouraged their scientists from publically debating the modern Creation scientists. Today, however, this debate continues through literature and other media published by Creation ('Young Earth Creation') science journals. These scientific journals and media compare results coming from the Evolutionary and 'Young Earth Creation' scientific models/methods.

In late 1980s, there re-emerged an earlier Theistic Evolution group under new names: *'Progressive Creation'* or *'Old Age Creation'*. Together with Neo-Evolution scientists, these *Hybrid Uniformitarians* began to compete with and challenge the original *'Young Earth Creation'* scientific interpretations.

After having examined these movements and many of their contradictory positions on scientific issues, one may ask several questions:

a) Had the original scientific method been improperly defined?
b) Had the method or the nature of science changed with time?
c) Wasn't the scientific method supposed to provide a constant - an authoritative anchor?
d) Has science and the scientific method been misapplied by some or all scientists?
e) Has science's content been misunderstood, blended or mutated into economics, philosophy or ideology?
f) Perhaps science and the scientific method had not been designed to fulfill a role that most thought it should fulfill.

g)   Has the scientific method expanded beyond its procedural scope?
h)   Does science now provide evidence for scientific misconduct or intellectual dishonesty?
i)   Have influential forces hijacked science and led to fulfill other purposes?

All answers to the above questions are in the affirmative. We actually find a mutating, speciating science that fights for legitimacy alongside with a healthy science. We can see this competitive spirit documented in scientific literature, textbooks and the media. Today's mutating science defines a truncated three to seven-step scientific method that we find in every textbook and on the internet. Philosophers of science provide additional scientific methodological steps and new definitions only to fail when counterproposals emerge. At the same time, we have witnessed during the 20th and 21st centuries, where political States and Empires interlaced their ideology with Darwinian concepts to help them achieve specific social engineering objectives.

When we listen to debates, read publications, observe the distribution of research grants; examine the *National Academy of Science* (NAS) rules and guidelines; scrutinize the education curriculum, education media and court decisions, it becomes clear that along with true science, totally new scientific terms and methods acquire new dimension of seeming legitimacy. This second type of science reflects what may be termed as 'scientificism.'

Comparison: Catholicism and Scientificism

Figure 1 compares '*scientificism*' and the historical notion of '*Catholicism*.' Each represents a practical approach to life, a knowledge base, a type of methodology that leads from uncertainty to conditional certainty, and is authoritative. Both reflect a purpose, objectives, strategy and a worldview – i.e., management.

Within Christendom – Eastern and Western Europe, Middle East and at Carthage, it was paramount to consider oneself as being a Catholic (universal) Christian. St. Ignatius first mentioned the concept in his letter to Christians in Smyrna in 106 AD (Ignatius, '*Letter to the Smyrnaeans*' 7 & 8,). This concept continued even after the schism (1054 AD), where both the Western Roman Catholic Church and Eastern Catholic Orthodox Church retained the concept. Martin Luther and other Protestants continued to see themselves as being Catholic for a time since an alternate position meant being less than Christian, a heretic, a revert to the paganism of ancient Rome, or one who entertained views of those who subscribed to the various pluralistic beliefs common among the surrounding non Christian tribes or nations.

By mid 19th century and specifically after WWI and WWII well financed 'scientificism' began to replace both the Catholic and post-Reformation views. On the one hand, *scientificism* seemingly promoted the scientific method, but also embellished it with notions of reductionism, uniformitarianism, modernism, behaviorism, secularism, dialectical materialist economics, and racial eugenics. Cryptically this movement also exerted concerted effort to replace Christianity in all social areas – specifically the notion of *Catholicism.*

Christians witnessed the early 20th century's dramatic events that engulfed the Christian civilization in two World Wars, revolutions, economic collapse, and devastating applications of radical scientific social sciences. All this made many Christians cautious, weary and skeptical about who controls the sciences. To respond or counter this skepticism, there emerged a new class of modern ideological scientists. They set out to sanitize the metamorphosing scientific applications that seemed to have been in the middle of all the social turmoil. These modern uniformitarian scientists purged the new method of all ideological derivatives. This approach was to render science as objective as possible – i.e., ethics-free. The new science and neo-atheism (e.g., R. Dawkins) were to obscure their contribution to these historical ideological and destructive excesses (over half a trillion casualties in Europe alone). Instead, the new 'science' was to project an objective value-neutral image of rationality.

Science definitions and terminology changed. Materialism became Naturalism. This approach reframed and repurposed all disciplines and subjects of science, letters and law. For example, evolution is now to be limited only to biology. Seemingly, paleontology and astronomy have been placed outside evolution's parameter (see NAS's Rules below). However, scientists still use *uniformitarian* (evolutionary) eyeglasses when they address non-biological subjects. Students learn 'science,' i.e., biology, physics, chemistry, astronomy, etc. but are not taught that these subjects may also seen in terms of, and within the framework of artistic, economic, agricultural, project, religious and management applications. Similarly, science policy-makers have also gone the extra mile to segregate science from alleged 'pseudo-science'. These ideological uniformitarians have demonized all that is 'anti-scientific': i.e., being religion, ideology, cults and race. Whole branches of knowledge have simply become less relevant.

It soon became apparent that there was more to these reductionist policies and notions of uniformitarian science. It was now possible to see within the emerging concept of Uniformitarian Scientificism that:

- In 1940 and 1995, *Scientificist CEOs* convened two key *'Scientificist Ecumenical Councils'* to formulate the *'Modern Evolutionary Synthesis'* creed.
- At the dawn of the 21st Century, the *Scientificist Magisterium* formulated the *'Encyclical of Infallibility.'* It proclaimed that: *'evolution is almost certainty, and is truth'.*
- During the past 30 years, *establishment scientificist organizations* had been screening scientists, administrators and personnel for their purity of 'scientific' thought, intent and practice. They unceremoniously defrocked, blacklisted and excommunicated numerous professional scholars, university professors, public

school teachers, government employees who succumbed to non-uniformitarian temptations.

- The *Scientificists*, during the past 30 years, with their trembling fingers identified an increasing number of heretics, backsliding and unrepentant reprobates among the '*Young Earth Creation*' and '*Intelligent Design*' scientists

- Some Federal State Courts tasked themselves to re-interpret the Articles of the American Constitution. Such re-interpretation now allows one to pass judgment on what constitutes true uniformitarian science vs. heretical cultist/religious pseudo-science. This re-interpretation sets ground for the excommunication of heretical scientists; prevent heretics from designing curriculum, publish science in uniformitarian scientific publishing houses, receive federal research grants, have access to repositories of scientific documentation and museum artifacts, prioritize scientific research in geologic, paleontological, oceanographic locations, and bar non-uniformitarian personnel from any institution that is under uniformitarian State authority. It is interesting to note, that the design of such decisions is similar to those uniformitarian courts that promote decisions on what constitutes 'hate laws,' 'rights for untraditional marriages' 'embryonic stem cell experimentation' and fetus abortion. In the recent past, counterparts to these uniformitarian courts and judges in foreign countries have legalized concentration camps, slave labor and genocide of hundreds of millions.

- Large sums of money support increased scientificist conversions. However, this initiative has taken unexpected turns, and remains an uncontrollably high-risk operation.

- Establishment scientists of all grades challenge and stir up the masses. These establishment scientists (commissars) work through public debates, publications, media and the internet. Their fingerprints are evident in every science education media programming.

- To establish its dogmas (purposes, objectives) and cannons (strategies), the Catholic Church has had thousands of years of historic precedents and the certainty of divine law. On the other hand, the *Scientificist Executives Magisterium* in spite of its world media outreach, digital networks, limitless financial support, their reductionist and modernist guidance has had only a little over a century. What kind of world would the Scientificists had created if they had thousands of years to produce their reductionist world?

- Scientists, who wish to see evolution as a scientific 'ethics-free' theory and an 'almost certainty and truth', increasingly, begin to realize that they cannot be scientists unless they address Scientificist ideological issues. These issues increasingly acquire ominous non-scientific content and form.

In contrast, it is possible to examine the two-millennial Eastern and Western Catholic Churches, which in spite of numerous major challenges had maintained a favorable and unique civilization until WWII. Historically, by 1000 A.D., the majority of European tribes and people converted to 'Catholic' Christianity. Christendom, in Europe stretched from the Scandinavian & Slavic Kievan Rus' in the East across to Britain in the West. At the same time, the few pagans who remained in the Christian midst initiated a counter Christian movement from the 14th century and accelerated a three-pronged momentum beginning in the 15th and 16th centuries.

It is, therefore, worth to compare a civilization that had been conducive to the development of 'accurate knowledge' (science) – Christendom - Catholicism, with one that has upheld relativistic priorities, reflects structures of failed civilizations and dark ages - Scientificism.

Figure 1 – Comparison: Catholicism and Scientificism

CATHOLICISM	SCIENTIFICISM
CERTAINTY	
Christianity has traced its certainty on legally historical precedents, established within and corroborated within the structure of geometric natural law. This historic documented legal content reveals three singularities: a) special creation; b) energy-density drop one cycle down the qualitative infrastructural cone; c) record of a recent global hydro-tectonic catastrophe, followed by environmental and biological adjustments.	Scientificism provides a synthetic construction of origins and history based on the doctrine of uniformitarianism: a) Current events, rates, processes are used to: b) Reconstruct history based on the principle of 'simple to complex'; c) Algebraic axioms, theorems, empirical materialism, methodological naturalism and reductionism. This creates a closed system (no external participation), whose origins may be a Big Bang or Steady State.
SCIENTIFIC METHOD	
Scientific method distinguishes between conditional certainty and uncertainty. It identifies the pursuit for accurate knowledgebase (science) thus laying the foundation for 'conditional certainty.' Providing investigative techniques, tools, processes to help resolve 'uncertainty' (decision-making and problem solving – DMPS – the scientific method.	Preconditioned filtering (5th step in DMPS) affects all decision-making and problem solving phases. This results in promoting 'naïve realism' through reductionism and modernism (uniformitarian ideology) and the 'Luxury Effect.'
AUTHORITY	
Two law-based holy/sacred traditions and Christian scriptures, which reflect: Ecumenical Councils (at least the initial 7 Councils that establish the original executive functions: purpose, objectives, strategies; and the supervisory functions: policy, procedures and rules. With the authority of bishops of the Church created a favorable environment not only for management style (ethics, attitude, culture – see Management Style in 3DMM) but also for the patronage and development of the	Rationalist and empiricist philosophies stand in contrast to, and enhance a replacement policy towards the Catholic option. Recently, there developed 'scientific' communities, establishment-sponsored organization, economics groups, marketing syndicates, university offices. Peer review, academic

arts, sciences, technology and commerce. This also included the development of centers of study, monasteries, universities and centers for the refinement of knowledge accuracy (e.g., Vatican library), through studies, scientific, medical and ethics groups under the Magisterium. Various Protestant authorities created environments for the development of sciences and arts. Continuously, scientific conferences & institutions, scientific editorial boards, publications, journals and newsletters help propagate information from the general knowledgebase.	councils seek 'overwhelming scientific consensuses.' Editorial boards on various publications, journals, newsletters; economic communities, 3rd party interest groups, State legal organizations, and international agencies.

To identify key components of these two sciences and scientific methods, it is necessary to go beneath the content of the current debate. An analyst must identify the underlying issues:

- Reflect upon *management dynamics*. This book selects one of the most reliable management models – the multi-billion dollar engineering project management. By summarizing this management model, it uncovers what drives science (knowledge base, 3DMM, 'conditional certainty').

- The *scientific method* contains the quantitative and qualitative stages (DMPS) that help convert 'uncertainty' into 'conditional certainty' (3DMM, knowledgebase. Below, the book examines three approaches to DMPS.

- Science and the scientific method have at least *12 limitations*. These in turn allow us to identify an *'Objective and Unfettered Scientific Method'* (OUSM).

- Being a sensitive tool, the scientific method contains at least *22 common pitfalls* that scientists, teachers and students must overcome. Also, recognize at least 20 common *pseudo-scientific methods* that compromise the DMPS process.

- Science and the scientific method cannot escape the *Five Models of Thinking* (mechanical, organic, processive and information-base models), and as a management tool, it must address *five qualitative levels of Change management*

We hear and read about the materialist/uniformitarian's definition of religion's unscientific foundations. Yet religion comes in many forms - cults, philosophies, theologies, and ideologies. The uniqueness of original Christianity is that it reflects geometric natural law. The contribution has led the 2000-year development of modern science and the scientific method. Many aspects of this Christianity have been 'forgotten' since World War I. Only a faint echo of this Christianity remains today. This book attempts to re-discover the millennial Christianity. This Christianity, founded upon geometric natural law contributed to scientific development.

The current task is to identify the true science and scientific method, and determine who among the four debaters – 'Evolution', 'Young Earth Creation', 'Intelligent Design' or the 'Hybrid Uniformitarians' comes closest to the 'objective or unfettered scientific method' (OUSM.). The 35 Figures & Tables provide a 'Scientific Score Card,' which helps identify who is most scientific.

CHAPTER 1 – INTERPRETATIONS OF SCIENTIFIC DATA

Today, education, the media, journalism and courts reflect and promote the uniformitarian interpretation of reality. Through this media, we also get uniformitarian interpretations of the Creation view. The pluralistic setting has re-prioritized history, linguistics and Christianity - now uprooted. Many religious organizations, denominations and churches have adapted too well to this new environment, changing their content while maintaining their original forms. This is why the historical Christian view presented in the two books may appear unfamiliar to most Judeo-Christians and Modernist Catholics. Yet, this original Christianity has and does provide the real alternative to the uniformitarian worldview. The 'Young Earth Creation' group represents many of these original Christian positions – legal history, recent creation, and three universal singularities.

Every culture has documented myths of origins. Early Greeks sought accurate knowledge, absorbed and developed many fields of inquiry that they inherited from previous civilizations. At the same time, the Greek myths provide a link to Greek iconographic artwork that represent biblical characters - Adam (Zeus); Cain (Hephaestos), Noah (Nereus), Nimrod (Hercules) and others.[vi]

Christian Biblical origins, however, are unique. This legal historical document contains features that are not evident in any other literature. The Book of Genesis, for example, contains the foundations of geometric natural law and geometric solids that reflects causes, processes, precedents, standards, rates, observations, measurable and testable conditions, and qualitative infrastructures. These features allow us to construction of workable knowledge models. Noah's Ark for example – the events prior to and during the Flood, the Ark's engineering and logistics design, allow us to reconstruct the Ark's ability to withstand unique global catastrophic conditions: accelerating oceanic tidal waves driven by lunar, solar gravitational and hydro-tectonic forces.

The Genesis description of seven-day creation and the global deluge events reflect engineering project management plans –e.g., the 'Gantt' scheduled activities, strategic resource loading and utilization, performance measures, forecasting, multiple infrastructural setups, quality control. The infrastructures align days and structures within the scope of geometric golden mean solids, energy density values, etc.

This same pattern is evident throughout the Christian Bible history. The 10 Commandments in the Old Testament (Exodus 20) are executive laws and not supervisory rules within the three-dimensional management model (3DMM). The

Tabernacle in the Wilderness encapsulates the Management Design infrastructure in the 3DMM (Exodus 25 to 40; also Hebrews 8ff), and King Solomon's Tabernacle (1 Chronicle 6) – reference Figure 5 and Tables 2 & 3 in this book. These Executive Laws further amplify the individual, family, group, social and global laws in terms of forecasted cause and effect 'blessings' and 'curses'(Deuteronomy 27-28) that have predictable sociological laws and effects. The New Testaments also identifies the 7-day Creation week where St Paul writes: 'the God who made the world and everything in it' (Acts 17:24-26); and that through Christ all things visible and invisible were made (Acts 14:15). Paul references the Genesis 1:1 – 2:4 account to identify Jesus Christ as the incarnate God of the Old Testament.

The recent 7-day creation of physical reality is evident throughout Christendom's history. This certainty professed from the beginning, documents events that occurred before and after the Flood, at the time of Mesopotamia, Babylon I & II, Egypt, Assyria and other civilizations. This Creation knowledgebase has withstood the competition that maintained views of eternal matter (steady state) (Greeks); evolution (Aristotle, Plato); cyclical expansion and contraction of the universe (India); mythologies that described a great universal battle between gods, and where the victorious god used the body parts of the defeated gods to create the existing universe (Babylonian Talmud).

In this competitive environment, the first Christian Church fathers asserted the recent Christian Biblical 6/7-day creation view and its recent creation. Justin Martyr, circa 100 A.D. mentions the 6-day creation in his 'Dialogue with Trypho.' Theophilus (169 A.D.) calculates 5,695 years since Creation in his 'To Autolycus.' Origen (185 A.D.) estimates that the world was 'not yet 10,000 years.' Eusebius (263 A.D.) mentions that several of the fathers wrote about a recent 6-day creation (Church History; Nicene and Post Nicene Fathers). Included is Basil (330 A.D.) mentions the recent six days of creation in his 'On the Hexameron.' Augustine (354 A.D.) suggests less than 6000 years in his 'The City of God.'

The history from Christian Biblical documents meticulously not only civilizations, events, correspondence, government protocol, but also parallel and serial genealogies, and many instances includes the party's number of years (age). This method of tracing people's ancestry had also been common throughout European history. They trace identities, specifically royalty to Noah's sons - Shem and Japheth. This was common among Christians as well as non-Christian nations, where for example Arthur Kostler in his book 'The Thirteenth Tribe' the Khazar aristocracy and the majority of its population that converted to Talmudic Phariseeism (10th century A.D.) documented their lineage to Japheth (son of Noah).

On the other hand, the *modern uniformitarian view* can be traced to the works of James Hutton's 'Theory of the Earth' (1785, and 2 volumes, 1795)[vii] where the author interprets geology in terms of 'older' geological formations. In 1811, Georges Cuvier and Alexandre Brogniart developed the 'stratgraphic succession' theory. The theory attempts to help explain Earth's geologic layers. Charles Lyell in his 'Principles of Geology Being an attempt to explain the former Change of the Earth's Surface by Reference to Causes now in Operation,' (1830) continued Hutton's and Cuvier's through and formulated the 'uniformitarian' concept for geology. Today this uniformitarian view suggests at least three to four concepts:

1. Start from today's conditions, processes, and rates
2. Through statistical progression, using the simple-to-complex (primitive-to-modern) process, (long ages that allow for statistical progression)
3. Use materialist and reductionist filters – closed system, and exclude open system that are designed to introduce external participating or creative forces
4. Affirm contemporary reality (#1)

Uniformitarians initiated these points from geology and paleontology. They universalized these points to other scientific branches. They then refocused the term 'evolution' on biology, but cryptically retained 'uniformitarianism' points for all other non-biologically related disciplines.

Scientists, writers, judges, politicians and specifically the debaters must be aware of the refocus, repurpose and shift approach. Examples, of biological refocus can be found as early as 1942, when Julian Huxley proposed an '*Evolutionary Synthesis*,' which more recently has been updated by Mayr, E. and W. B. Provine, eds '*The Evolutionary Synthesis: Perspectives on the Unification of Biology*.' This last work lists a five-point strategy for uniting biological subjects under an evolutionary synthesis. He thus provides criteria or assumptions for constructing the evolutionary scientific model.

Similarly, uniformitarians are active across every field or reality. To better, understand the *Modern Creation Science* initiative within this uniformitarian environment - historical and social content – it is important to understand the events that occurred during the early part of the 20th century in the USA.

• During the past 250 years, uniformitarian views aimed at reducing certainty via relativism. This has existed in education and seminary curricula in Europe and America. Uniformitarians reinterpret every domain: history, geological, paleontological, archaeological evidence, linguistics, anthropology, oral traditions, economics and specifically those that focused on the various criticisms of Biblical interpretations. Here, during the past 100 years, schools of thought provide an uniformitarian slant on every subject – e.g., source criticism, redaction, form, canonical, rhetorical, narrative, psychological, socio-scientific, and higher and textual criticism, and the modernism and post-modernism. The modernists aimed primarily the Roman Catholic sacred traditions, scriptures and authority.

All of these criticisms, which appear logical and legitimate scholarship, actually become ideological powerhouses that specifically aimed to reduce Christian certainty (knowledgebase, 3DMM) through relativism. All curricula in Europe and America had been 'updated' to meet the new objectives. Although this uniformitarian initiative succeeded in Europe, in the USA this alarmed conservative members of many Protestant denominations. 'Fundamentalists' detected a clear subjective anti-Christian secular agenda, akin to that deployed in the USSR, and in the 30's in Nazi Germany. The conservatives identified common links to:

• Darwinian notions of 'survival-of-the-fittest,' that debased the European civilization's moral fiber. This contributed to a self-centered and irrational momentum that led to excesses during WWI: galvanized nations; promoted various spectrums of Socialism – economic foundations, justifications for collectivism and behavioral sciences.

• In America, among other things, these tendencies led to the 'Pre-Conference on the Fundamentals of Our Baptist Faith' (June 21 & 22, 1920) at Delaware Avenue Church in Buffalo, New York. It is here that the original Baptist Biblical fundamentals were established – including scriptural inerrancy. While the period that lead up to WWII, in Europe, even though Darwinism had been accepted and even promoted by many of the protestant institutions, the pre-Vatican II Catholic Church and its Continental, European-based Creation scientists, understood the implications of uniformitarianism. The Catholic Church sanctioned Teilhard the Chardin's evolutionary concepts that allegedly led to an 'Omega Point'; and issued encyclicals against 'Modernist' activities – see the First Vatican Council's – 'Humani Generis'; Pope St Pius X's, 'Pascendi Domenici Gregis (Encyclical 'On Modernism'), par. 10.[viii]

• In 1925 uniformitarians of various persuasions found an opportunity to act against the religious fundamentalist counter-initiative. Uniformitarians took advantage of the '*Scopes Trial*' in Dayton, Tennessee. The initial issue on trial (the Butler Act) expressed concern for teaching notions of the 'struggle for existence' and that 'man has evolved from primates.' These two notions had proved to have an adverse effect on moral issues. Undercurrents during court proceedings, however, had helped sway national and international opinion. Henry Louis Mencken, journalist of the *Baltimore Sun* daily provided analyses of trail proceedings and forced alternate court agenda. Mencken's reports misrepresented court proceedings, events that surrounded the trial, and the roles of the various players and personalities who participated in the trial. Through this, Mencken totally distorted standard practices of objective reporting of court proceedings.

When someone examines actual court proceedings and the surrounding events, one finds that Mencken established a precedent for modern uniformitarian journalism and American reality. Mencken formulated the competitive view between materialism (science) vs. myth (religion) that influenced modern media, school textbooks and court proceedings. In 1925, national and international newspapers (media) transmitted whatever the '*Baltimore Sun*' reported on the '*Scopes Trail*' - new views that evolution is science, enlightenment, and the hope of all future progress, while religion, and specifically Christianity, is prone for bigotry, cultism, backwardness and reflect dark age pseudo-science.

From 1925 to the 21ˢᵗ century, the *Baltimore Sun's* strategy found itself expressed in countless plays, movies, curricula and court decisions. None of the writers, authors, producers and judges cared to review the actual documented court proceedings. The *Baltimore Sun's* 'social realism' perspective is the current view of what had transpired during the '*Scopes Trial*' and what is what constitutes the authorized American**ist** view

The only scientific evidence brought during the *Scopes trials*, was a tooth of a newly discovered 'Nebraska Man'. This was to be scientific evidence for evolution. After the trial, scientists traced the tooth to its original location. Scientists discovered that the tooth was not that of a humanoid but of an extinct pig.

'*Scopes Trial*' events led to several undeniable conclusions. That:

1   The 'scientific evolution' movement was a highly financed program that had strategic, tactical and covert objectives. These uniformitarian objectives were to undermine the original American Constitutional and Christian foundations of America. A similar program had successfully implemented in Europe.
2   Increasingly, establishment scientists are to staff science labs and administrative posts. They are to approve and authorize all stages of empirical evidence. They will determine the distinction between what is scientific and pseudo-scientific
3   They will staff ideological, social and legal activists' movements who market, defend and are willing to go to extremes in order to ensure the dominance of the uniformitarian worldview and system.
4   A governing entity supports and authorizes the content of an exclusive and comprehensive educational curriculum. This entity issues only that which promotes uniformitarian premises and objectives.
5   A financed and controlled media guides, promotes and saturates the audience with an exclusively uniformitarian outlook
6   Policies of pluralism maintain a substitution and replacement program for original Christianity
7   A social re-engineering program conducted within the scope of uniformitarian 'secular' initiatives affects all social organizations.

It is not until the 1950s that features of the 100-year old unobservable 'evolutionary synthesis' had actually been tested in the scientific lab for the first time. It is also within this context that modern Creation science was born. Until this modern period, many European and American science writers, scientists and American fundamentalist science writers formed the Creation and the counter-evolution organizations. In 1961, however, Dr. Henry M. Morris and John C. Whitcomb co-authored the book '*The*

Genesis Flood,' (Institute of Creation Research, CA 1961) and launched the modern Creation science. Dr. Morris (1918 – 2006) provided a *'Creation Scientific Model'* to help interpret and predict the voluminous scientific data. He published and debated evolution scientist contrasting the results of both the Creation vs. the Evolution Scientific Models. The Creation Scientific Model tested geological and paleontological evidence and proved that the results provided and predicted a better explanation than the Evolution Scientific Model.

It became evident that the Bible was not only a document of legends, fables and moral teachings. Instead, it became clear that the Bible is also a unique source of information that documents verifiable historical events; legally documents a 4000-year history and through this provides verifiable and reconstructable scientific evidence, knowledge, standards and observations. Henry Morris demonstrated that when interpreting the original historical Biblical text in non-uniformitarian terms, the legal Biblical history provides significant and key verifiable evidence for a recent worldwide Flood - geology, paleontology and other scientific disciplines. During the 1970's Henry M. Morris and his colleague Dr. Duane T. Gish participated in numerous middle and high-level public debates with scientists who supported the evolutionary view. During this time, Morris and Gish had succeeded in presenting stronger and more convincing arguments and evidence for the creation science model than that presented with the evolutionary scientific model.

1.1 – Comparison of Two Basic Scientific Models

As listed above, *Evolution* scientists construct, interpret and predict scientific data from the Evolution Scientific Model - at least four uniformitarian assumptions:

1   Start from today's conditions, processes, and rates
2   Through statistical progression, using the simple-to-complex (primitive-to-modern) process, (long ages that allow for statistical progression)
3   Use materialist and reductionist filters – closed system, and exclude open system that are designed to introduce external participating or creative forces
4   Affirm contemporary reality (#1)

These are *ideological* positions and not scientific ones. These axioms or theorems have no objective (scientific) reference or proof (see Figures 19 and 23 for more detail). In other words, instead of prioritizing and relying on actual historical evidence, evolution scientists attempt to re-constitute a synthetic past and a history of origins on these four uniformitarian doctrinal assumptions.

Creation scientists construct, interpret and predict scientific data from the Creation scientific model also recognize at least four conditions (note: Creation scientists are those who existed throughout Christendom, and can include, the modern ones in the USA.)

1. Begin with current o millennial historical evidence as documented and corroborated with the legal history of the Christian Bible
2. Use existing empirical scientific evidence (all branches), discoverable physical natural laws (e.g., motion, gravitation, biogenesis, geometric); biological complexity as it adapts to changing and challenging environments. Recognize the distinctions among five change levels: copy, adaptation, re-engineering, re-design, re-invention
3. Consider the documented record(a) that identifies three recent historical singularities – specifically the third for which corresponds with supporting empirical evidence
4. Using the Creation Scientific Model, reconstruct conditions and events that occurred in the past; and forecast how these help explain current conditions.

1) Singularity 1: A recent special creation (self-identifying Creator, Eternal Lord God, Prime covenant designer) - intellectually recognized through *geometric natural law* (see Figures 20, 24, 26). The divine creative process follows a specific timeline - documented from the Creator's perspective (Gen 1:1 – 2:3), beginning with the creation of time itself (Gen 1:1) – see Figure 26 for details.

2) Singularity 2: A drop of at least one energy-density cycle on the universal infrastructural cone (see Figure 22) – within three days after the creation week

3) Singularity 3: Within less than two millennia after Singularity #2, events of an initial global hydro-tectonic catastrophe of relative short duration (12.01 months) that was followed by secondary multiple mega-environmental adjustments of longer duration. This occurred in areas of the atmosphere, weather, geology/tectonics, sea levels, volcanic activity, a *circa* 500-year ice age, creature adaptation to new environments. These Singularity 3 conditions and effects are subject to scientific investigations - observation, testing and prediction through branches of science - geology, paleontology and other branches.

The purpose for scientific models (Evolution, Creation, and Intelligent Design and Hybrid Uniformitarian) is to help interpret data, make predictions and determine which of these represent a better scientific model.

Figure 2 - Comparison: Uniformitarian vs. Creation Scientific Predictions

UNIFORMITARIAN / EVOLUTION		CREATION SCIENCE
ASSUMPTIONS		
Scientific predictions derived from uniformitarian, materialistic, reductionist constructed past and present rules for interpreting empirical data.		Three historically documented singularities, geometric natural law that account for life's original complexity. while the third singularity is evident through empirical scientific methods
TIME		
Prediction: slow	Prediction: Creation events within the context of geometric	

random/chance process requires billions of years to bring the Big Bang events to present conditions. Prediction: time/age of fossils measured via geologic column, which results from the slow processes of localized sedimentary or episodic catastrophic events (e.g., meteoric impact). Prediction: simple creatures found at lower sedimentary layers, and species that are *more complex* found at increasingly higher layers. Prediction: various chemical dating methods can ascertain long ages	natural law identifies the creation of time and provides a timeline record for all creation, three universal singularities that affect directly and indirectly all temporal estimations of age and provide evidence for recent age of Creation between 6000 to 12,400 years. (John Morris, '*The Young Earth: The Real History of the Earth – Past, Present and Future,*' Master Books, AR, 2007, p. 35) Prediction: the geologic timeline (column) reflects not long uniformitarian ages but the effects of sedimentation occurring during an initial global hydro-tectonic mega-catastrophic event, with subsequent secondary self-adjusting catastrophes over hundreds of years. Under such conditions geologic sedimentary layers and the fossilization of living creatures' mobility capacities as they attempt to escape flood conditions – slower creatures will be buried first (at lower strata), and faster ones under sedimentary layers at increasingly higher layers. Geologic sedimentary formations are a record of the speed of creatures relative to their environmental conditions, rather than the age of the fossilized creatures. Prediction: significant inconsistencies in the chemical dating methods, that compromise every dating method, particularly when these methods are compared with each other.

GEOLOGY AND PALEONTOLOGY

Prediction: geologic column and the fossil record reveal fossilized life forms that developed over generations from simple to complex structures – net basic increase in complexity over time, with unlimited vertical change, with unlimited vertical change. Prediction: simple to complex fossilized life forms found in areas around the world. Prediction: voluminous amounts of intermediate / transitional forms leading from one species to the next will be evident.	Prediction: geologic strata, sedimentation and the fossil record reveal recent events of an initial global hydro-tectonic catastrophe and secondary environmentally self-adjusting mega-catastrophes: evident in geology, paleontology, hydrography, atmosphere, ice age, demographics, etc. Prediction: geologic and paleontological evidence will reveal that life appeared fully formed (e.g., difference between the so-called pre-/Cambrian age). Clear and sharp boundaries will separate original taxonomic groups. Prediction: There will be a total absence of intermediate/ transitional forms in the earth strata. Prediction: Evidence of polystrate tree trunk positions (allegedly standing across billions of years) are the result of rapid sedimentation under hydro-tectonic effects – this rapid process is currently observed in the events following mount St. Helens volcanic effects. Prediction: identified symptoms of 'gigantism' among plant, animal and human fossils when geologic layers reveal specific effects of the initial mega-hydro-tectonic effect.

MICRO & MACRO CHANGE MECHANISMS

Prediction: universal common ancestry extending from atom to living cell and through various branching of the standard phylogenetic tree to the emergence of man. Prediction: although evolutionary change is too small to observe, measure, predict and reproduce, it can be used to estimate that random beneficial mutations among groups in challenging	Prediction: the created original kinds/ syngameons 'species' will vary due to environmental challenges but speciation will be limited within each kind / species. The taxonomical tree simply identifies groups and families, not their change from one species into another. Prediction: 'beneficial' random mutations must overcome the significantly greater volume of hazardous mutations. An organism's security programs help prevent or remove anomalous mutations. Biological system is programmed for adaptive change (micro-evolution), but not for higher infrastructural change that involves re-engineering of supervisory systems and re-design (i.e., macro-evolution)

environments will allow groups of species to not only to adapt (micro-evolution) to their environment, but also change their species (speciation – macro-evolution) over time and environment. Prediction: appearance of new species demonstrates new and addition of qualitative information. Prediction: the appearance of greater variety of new species.	unless such 'micro-change' has been programmed into the genetic code (e.g., caterpillar to butterfly). Prediction: any significant environmentally challenged adaptive change demonstrates a loss and not a gain of information. All of today's species are descendants of a few original species 'proto-types' - syngameons Prediction: evidence shows for the high potential of gradual extinction of species rather than their rise in complexity.
LIFE	
Prediction: the axiomatic concept of abiogenesis suggests that life originated from non-life.	Prediction: Evidence of biogenesis – is a law that shows that life comes from life. Paleontological evidence shows that all life appears fully formed, and then adapts within the scope of its programmed structure.

1.2 - Hybrid Uniformitarian and the Intelligent Design Views

The *hybrid uniformitarian* views include such theories as:

1) *Theistic Evolution* - God used evolution to create the world
2) *Progressive Evolution* or *Long-age Creation* - Creation days were actually long uniformitarian ages but the timeline is 'punctuated' by divine interventions at critical points to introduce qualitative new designs. This would account for gaps in the fossil record, and absence of transitional forms
3) *Day-Age* - the Biblical days correspond to long ages
4) *Gap Theory* - billions of years existed between Genesis 1:1 and 1:2
5) *Framework Interpretation* - known as a 'literary framework' the six Genesis days were not literal, not scientific, it is just a religious doctrine of creation.

This book does not focus on hybrid uniformitarian views because such models must reflect key 'scientific' concepts. A summary of the Hybrid Uniformitarian synthetic constraints and artificial means:

1) Consider irreconcilable notions of uniformitarian and creation *origins* - a timeline difference suggested by evolution (long ages) and Biblical Creation (recent ages). This is particularly true when one compares the sequence of 'creative' events (see Figure 3). There is a clear mismatch among the events. Many events reflect reverse action. Where the Creation model reflects geometric natural law (Figure 21 and 24) and geometric solids (Figure 21), the evolution model has none of these.

Figure 3 - Comparison: Origins Interpretation – Evolution and Creation

From: John Morris, '*The Young Earth: The Real History of the Earth – Past, Present, and Future*,' Master Books, AR, 2007, p. 31 [the right and left column reversed]

EVOLUTIONARY ORDER OF APPEARANCE	BIBLICAL ORDER OF APPEARANCE
Matter existed in the beginning	Matter created by God in the beginning
Sun and stars before the earth	Earth before the sun and stars
Land before the oceans	Oceans before the land
Sun, earth's first light	Light before the sun
Atmosphere above a water layer	Atmosphere between two water layers
Marine organisms, first life form	Land plants, first life forms created
Fish before fruit trees	Fruit trees before fish
Insects before fish	Fish before insects
Sun before land plants	Land vegetation before sun
Land mammals before marine mammals	Marine mammals before land mammals
Reptiles before birds	Birds before land reptiles
Death, necessary antecedent of man	Man, the cause of death

2) Hybrid Uniformitarians needlessly distort or dilute the uniformitarian and special creation models by re-interpreting and proposing some external 'dynamic' physical processes that cannot substantiated neither on uniformitarian nor on special creation grounds. Hybrid notions cannot explain the specific external interventions through reasonable and evidential proofs, nor on Biblical scriptural evidence. Hybrid Uniformitarians do not substantiate the long vs. recent ages of the Earth or the Universe.

3) Hybrid Uniformitarians resort to 'leaps of faith' – i.e., believe in certain conditions that cannot be deduced from either the competing standard models or from a systematic treatment of evidence. It is important to note that although the notion of 'leaps of faith' is usually associated with religious practice, 'leaps of faith' is evident among those who propose materialistic explanations. For example, in the absence of concrete scientific evidence, evolutionists refer to their exclusive 'consensus of scientific opinion.' Mayr proposes in his 'The Modern Evolutionary Synthesis' that scientists extrapolate micro- to macro-evolution, with historical contingency, which are to be defined as explanations for different levels, since gradualism does not mean constant rate of change. Here explanation and extrapolations must substitute empirical evidence for something that may/does not exist and are unprovable.

4) Hybrid Uniformitarians complicate or do not resolve definitional and evidential parameters.

5) Hybrid Uniformitarians provide a synthesis, which instead of reconciling, actually places all parties at odds with each other.

6) Hybrid Uniformitarians re-define and re-interpret: a) the original 'theological' foundations that had been established over the millennia (Christianity) and in many cases inadvertently offer pantheism as a solution; and b) re-define and re-interpret the original uniformitarian view that has been established on purely materialistic and

reductionist principles. The latter approach do not allow for metaphysical participation or speculation.

1.3 - Intelligent Design

There exists a difference between the standard Hybrid positions and methods proposed by the 'Intelligent Design' scientists. ID scientists propose a Teleological approach to help explain totally unexpected and unpredictable scientific evidence at the sub-chromosome level. The Intelligent Design group formulates a hypothetical, eclectic and all inclusive notion of an 'Intelligent Designer' in order to avoid falling into the usual trap that many neo-evolutionists have fallen into - where in the absence of predicted data these scientists improvise and introduce notions of 'hopeful monsters,' 'punctuated equilibrium' and hypothesize about 'extraterrestrial' participation or intrusions into the evolutionary process.

 The notion of an 'Intelligent Designer' is a hypothetical and inevitably a pantheistic view that can account for the discovered super high information technology and nano-robot (nanobot) activity and processes evident that the sub-chromosome level. It is not an attempt to introduce elements of organized religion, e.g., the Christian Biblical God, who has a name, purpose, historical mission, social and inheritance laws that require a 'priesthood' for interpretation through some revealed communication. The Intelligent Designer is an impersonal, non-theological entity and is not described within a creative timeframe.

Intelligent Design scientists have their own personal views on the identity and activity of this Designer. I.D. scientists simply wish to relate that no amount of billions, trillions or zillions of years could account for a statistical explanation for the complexity observed and recorded at sub-chromosomal information levels. However, Creation scientists who look at the Intelligent Designer – recognize, within the scientific framework, some general features and effects of a participating and interacting supernatural being. Also, recognize the pantheistic framework of this scientific hypothesis.

Within the Creation Scientific framework, the divine Creator is revealed through: a) legal historical documented evidence, b) geometric natural law; and c) physical evidence within Creation that can be observed, documented, tested and verified through the original definition of the scientific method, which helps ensure an accurate knowledgebase. Intelligent Design scientists come to their conclusion about the existence of a hypothetical Intelligent Designer after the fact (examination of scientific evidence) – *a posteriori* approach. The Creation scientists have the framework before all the scientific evidence is tabulated – the *a priori* approach. Curiously, the uniformitarian evolutionists also have *a priority* approach - matter.

1.4 – Three Debaters Formulae

The three remaining debaters' formulae identify key points that the model represents:

The basic <u>neo-evolutionary scientific methodology</u> summary is:

$$\frac{C + M + T + V}{L + E + R} = A + K + S + P$$

C = Chance (e.g., genetic mutation, flow, mechanisms)
M = Matter
T = Time (current time)
V = Environmental stress
E = Uniformitarianism
L = Relativism
R = Reductionism
A = Simple (primitive) to complex (natural selection, across 5 infrastructural change levels; micro-/macro-evolution
K = Knowledgebase (materialism, uniformitarianism, reductionism, relativism)
S = Standards
P = Scientific Process (evolutionary realism, dialectical materialism)

The basic <u>Creation scientific methodological</u> formula is:

$$\frac{C + M + T + I + V}{L + E + R} = A + K + S + P$$

C = 3 historical singularities: special creation, universal energy-density drop by one cycle on the infrastructural cone; one super global hydro-tectonic catastrophe followed by secondary mega global adjustments (environmental Stress = 'T').
M = Matter
T = Time – affected by 3 singularities (see 'C') and Recorded ('R') in legal documentation.
I = 5 infrastructural change levels designed in living organisms
V = Environmental stress
L = Geometric Natural Law
E = Authoritative legal historic documentation
R = Geologic and paleontological evidence and predictions
A = Adaptation to environment (2 infrastructural change capabilities to adapt to the environment – copy and adaptation)
K = Knowledgebase (certainty)
S = Standards, codes (certainty)
P = Verifiable lawful scientific process to resolve 'uncertainty' DMPS

The basic <u>Intelligent Design methodology</u> formula as:

$$C + (T - E) + M + I + N = A + K + S + P$$

C = 1 singularity: special creation – Intelligent Designer
T = Current and Laboratory time
E = Evolutionary Time to prove statistical progression [subtracted]
M = Matter
I = Nano-level information engineering; harmonized laws of the universe.
N = Nano-level robotics engineering
A = Irreducible complexity with unfettered scientific methodology
K = Knowledgebase (certainty)
S = Standards, codes (certainty)
P = Verifiable lawful scientific process to resolve 'uncertainty'

For example, how differently do the debaters address the **'time' variable ('T')**? Each treats time in a different way.

Neo Evolutionists use the uniformitarian platform to reconstruct the past (and future) upon which all existing physical conditions and forces exist. In other words – 'use the present to measure the past'. To accomplish this we have seen up to four steps, here are the common three steps:

a)   Using current standards rates and processes
b)   Extrapolate physical conditions backwards from present to past to the primordial point, and then allow statistical progression (chance) from simple to complex to current conditions.
c)   Using statistical progression under challenging environmental conditions, have single living cells move through various intermediary species to man.

The 3-phased view theorizes uniformitarian time. In other words uniformitarian time is the byproduct of chance (statistical progression) allowing it to work itself in a challenging environment (mutations) from matter to living cell to man.

Creation Science time is associated with the recorded legal history of geometric natural law and the three singularities: special creation, down scaling of energy density, and the hydro-tectonic initial and secondary self-adjusting catastrophes. We can deduce this time from three sources: a) interpretation of legal historic documentation; b) effects within natural physical environments (geology, paleontology, etc.); and c) corroborated through contingent scientific evidence and predictions. Each of the singularities affects time differently. This affects dating methods, rendering all geologic and cosmic conditions to reflect significantly more recent periods.

Intelligent Design - the individual ID scientist may hold private views on time. They, however, will negate in one voice the Neo Evolutionary billions of years when these years are used to suggest statistical possibility or probability for the development of nano-information engineering complexities at the chromosome level. The 19th century Darwinian Evolution constructed on an organic model of thinking cannot accommodate information technology modeling. The Intelligent Design scientists, therefore, do not offer any official direct methods for establishing periods (historical). They recognize that no amount of time can account for the nano-information engineering complexities, and this warrants a new theory.

CHAPTER 2 - QUEST FOR CERTAINTY (3DMM) AND SOLUTIONS FOR UNCERTAINTY (DMPS)

Modern management theories consider knowledge as a resource. Companies, corporations and nations in competitive environments consider knowledge in terms of a potential strategic advantage that helps reduce risks and hazards. Knowledge is 'intellectual capital' that is manageable. Knowledge Management (KM) users can conceive, define, verify, validate, archive, outdate, innovate, distribute to single, group and networked users in timely and usable manner. Knowledge helps improve learning, performance, research (e.g., engineering, marketing, education) and promote general awareness.

Knowledge when codified and framed in the form of expert systems will contain specific procedures that assist in decision-making and problem solving. Knowledge systems allow users to simulate or convert inputs, practices or criteria (e.g., medical, financial, operating, legal) into usable products. Similarly, there are techniques and methods to simulate and interpret any body of knowledge.

Engineering has gone the extra mile and budgeted to ensure the fulfillment of plans' objectives through qualitative information. Modern biologists identify information software and robotics processes at micro-nano levels within each living cell. They thus detect, define and measure evidence for pre-programmed networks and processes that will not be managed in our scientific labs for at least 500 years. These nano-level processes suggest the existence of built-in communication networks, infrastructures, supervisory and executive systems, and new standards that form the core and essence of the knowledgebase. Knowledge, at certain levels, can operate within supervisory systems (automated, instinct), as well as, in the case of humans at the conscious, awareness and executive levels. Scientists and engineers can understand this within the 3DMM – which contains infrastructures, executive, supervisory and functional systems.

Knowledge-seekers and consultants explore issues of knowledge management. The ancient Greeks had developed intellectual structures and mathematical processes that helped them define and arrange data, information and knowledge in a manageable manner. Mythology also reflects a knowledgebase, which together with economics become a social vehicle that navigates with a purpose through areas of productivity.

This vehicle carries not only knowledge but also conditional certainty in an attempt to meet objectives and purposes.

Some have devised such vehicles to help understand nature, God's will or God's nature. Aristotle views this vehicle intellectually. Aquinas offers cosmological arguments – 'first cause' and the 'goodness of God.' St Anselm provides an ontological explanation by presenting a range of Creation's good features. This has also been evident in mathematics, geometry, constructions of pyramids, megaliths and the Zodiac. The Renaissance emphased humanism, accuracy of knowledge and development of scientific branches – see the works of Erasmus, Francis Bacon, Galileo Galilei, to Riemann.

Following the ravages of the 16th and 17th century religious wars, an intellectual movement set itself the task to re-examine the certainty in religious revelations. In the 16th century, Rene Descartes began a rationalist quest for certainty. His approach was to question the primary sources of knowledge. He introduced extreme skepticism by elevating doubting above everything except his own subjective ability to think - 'I am therefore I am'. This Rationalist philosophy was set upon self-evident axioms (Baruch Spinoza), suggested a pantheistic dualism through the unity of nature and God (Gottfried Leibniz). Later, Immanuel Kant attempted to organize philosophy on rational, skeptical, logical and axiomatic grounds. In the meantime, as Rationalists worked in France, Empiricists made some counter offers in England. Empiricists discounted these innate ideas (Rationalism) and proposed that one can achieve certainty only through observation (John Locke, George Berkeley, David Hume).

Together, Rationalists and Empiricists contributed to the foundation of the Enlightenment. Enlightenment philosophers achieved their enlightenment through their emphasis on nature's order, rigor and reductionism. From this, they derived and inferred many other ideas such as freedom from dogma, organization of States into self-governing republics via democracy, religious tolerance, scientific method, market mechanisms, capitalism and reason as being the primary value of society, the freedom to pursue truth without sanction for violating established ideas).

Throughout this 18th-century 'Age of the Enlightenment', there was a systematic search for pure empiricism (natural philosophy). Newton substantiated this empiricism, and was followed by Diderot, Voltaire, Rousseau, Montesquieu, and Kant.

The 19th century's concept of '*self-organization*' and '*intrinsic order*' was launched through Kantian metaphysics, and continued in the form of a Hegelian dialectical process. Here, knowledge and reality automatically ordered itself (self-regulated) through inherent organic dynamic forces - thesis-antitheses-synthesis. Hegel philosophy influenced a stream of philosophers, scientists, economists and social theorists.

Just as the Enlightenment and the Encyclopaediists overturned well-established traditions; 20th century Modernism reset Enlightenment and rationalist-based philosophies. Modernists rejected Enlightenment's foundations of knowledge and certainties, by introducing a radicalized reductionism. Modernists refuted irrationality and emotionalism and promoted a new social economy. Based on Hegel and Furerbach, Modernists laid the foundations for understanding materialism's link to the science of economics (Karl Marx's 'Das Kapital'). In other words, here science was nothing more than economics. Others began to postulate on this new empiricism. Freud turned psychology upside down, while Friedrich Nietzsche analyzed and rejected religion, philosophy, morality, culture and science itself. All of this also led to a post-modern re-design and re-interpretation of all the benefits that the Age of Enlightenment had brought. It became apparent from the mid 20th century that certain features of the Enlightenment had become a liability. After all, Enlightenment suggested a breakup of reality into specializations, while ignoring traditional wisdom and its potential lateral consequences. Instead of idealizing enlightenment figures such as the founding fathers of the United States the art of Reductionism that began with Descartes [Part V of his Discourses (1623)] and blossomed in Positivism [begun with Auguste Comte (19th century)], in the latter part of the 20th century philosophical thinkers created their own definition of positivism (Emile Hennequin, Wilhelm Scherer, Dimitri Pisarev, Stephen Hawking). They associated authentic scientific knowledge with modern positivism and made it synonymous with empirical reductionism.

The rational-empirical-positivist path relies on mechanistic, organic and processive paths that are within a closed system. These closed systems are independent of any external influences or participation. The intellectual part of the mind finds comfort in the freedom from external dogmas, laws and objective frameworks. Yet, there were those who still attempted to introduce open systems (divine concepts). They dressed such attempts in rationalist/empiricist terminology and concepts but, at best produced pantheistic results.

By definition, the intellect, along with its workings, perceptions and interpretations – has temporal features. In this view, this approach would inevitably tend to reject anything that would emanate from beyond normal human perception. As David Hume (1711-1776) suggested that there is a rift between the 'what is' and the 'ought to be,' ('scientific relativism'). This unbridgeable gap can be breached artificially - through hypothetical (relative) 'leaps of faith' (subjectivism). Such artificial bridging subjectivism would express itself in terms of a: myth, ideology, religion, and even through the fifth point (thesis) in the 'Modern Synthesis' – where an 'explanation' substitutes scientific evidence.

The intellectual positivist and reductionist premises that underlie the uniformitarian concept reflect:

a) Mechanical, organic and processive models of thinking (see Figure 15) forced to be exclusivist, supremacist and pan-/poly-theistic because it rejects all other legitimate options (e.g., information-based modeling and more specifically the geometric models of thinking).

b) An '*a-priori*' position, which prioritizes materialism and defines all reality in terms of de-prioritized values – irrational, mythical, cultist, subjective, emotional and religious.

c) Because of (a) and (b), an axiomatic Uniformitarian Realism (UR). Moreover, without realizing it, this view falls into the scope of uncertainty that cannot deliver levels of guaranteed, quality or contingent certainty.

The atheist, for example, must *first* rationally recognize, formulate and define the existence of some divine force, being or entity. Usually this is defined in terms of mythological, ideological, and religious terms that inevitably have a pantheistic scope. This inevitability arises out of the rational construct. *Second*, the atheist must then negate this pantheistic construct by filtering such reasoning through skeptical or materialistic axioms and theorems. *Third*, the atheist must convince the audience that this materialistic model of definitions and deductions applies to all open-system views that propose non-materialistic causes and participations -see Rule 22 Chapter 3.3 below for further details.

Such reasoning and definitions derive from the human supervisory application (D4 – see Figure 5) and exclude executive faculties that deal with purposes and laws (see Figures 9 through 13). However, matter is not the sole domain of the materialist. One does not need to be an atheist and skeptic to recognize the value, reality and the economics of matter. All ancient civilizations, as well as modern theologians and scientists expressed belief and their understanding of the material phenomenon and its laws. The Christian Bible, for example, clearly indicates the multi-nature of matter as a product of a Creative/ Designing Supernatural Being. In other words, atheistic, creation and intelligent design (teleological) terms and concepts can easily interpret the economy of 'matter.' The only difference is that the atheists, skeptic and uniformitarians see matter in an *a-priori* fashion - as a beginning and concluding concept (closed system), whereas we can conceive matter in terms of geometric natural law – i.e. a means to an end.

Some skeptics have correctly suggested that the seeming orderly makeup of the universe actually contains too many defects, errors, mistakes and inconsistencies. And suggest that such imperfection is clearly inconsistent with the attributes of a perfect Creator designer. However, such critics, on the one hand, become lenient and allow for a rationally formulated god (pantheism or polytheism) to create an imperfect world (Hybrids). At the same time, such critics would not consider that the existing universe could be traced to causes and conditions that are beyond what the rationalist filtering process. Specifically, to documented legal history that identifies an original perfect creation that succumbed to two global singularities: 1) a drop of at least one infrastructural energy density – thus resulting in Creation's 'starvation' for energy; and 2) the global hydro-tectonic catastrophe whose evidence can be read in the Earth's geologic record. The original perfect creation (1st singularity) existed prior to the introduction of the following two singularities. This original perfect creation was dynamic enough to allow adaptation to new levels of imperfect economies due in large part to the dynamics of geometric natural law. This Creation has retained intricacies, up to five levels of complexity and esthetic value as interpreted by the human eye and scientific instruments.

This brief outline, identified philosophers, economists, linguists and scientists who attempted to identify the source and nature of certainty achieved through intellect alone. This endeavor, however, left many questionable methodologies that the following thinkers rejected, or methods that dwindled to simple word games, symbol juggling, axioms, theorems, postulates, value relativism and concepts that simply left a muddy dirt road of Uncertainty.

The millennial search for knowledge and certainty may easily be resolved by recognizing that all of these systems reflect manageable information. Both information and knowledge reflect manageable principles. This history of ideas, viewed from a management perspective, reveals knowledgebase management principles that are by necessity commonly used in business, engineering projects and corporations. The strengths, weaknesses, value and ideas become easy to understand within a managerial environment. Similarly, philosophy, myths, ideologies, and the scientific method itself are nothing more than manageable information. This managed information reflects several infrastructures of conditional certainty, whose uncertainty must be processed through the DMPS to achieve a greater conditional certainty within the 3-DMM.

It is challenging in our information-saturated age to find a comprehensive description of the scientific method. The 'method' recognizably is a strategy, technique, procedure, or set of rules for conducting observations, presenting alternatives that must be filtered for testing and implementation. Etymologically the adjective 'scientific' can be traced to the Latin scientia 'knowledge' (from scio – 'I know'); while the Indo-European, Sanskrit, Greek terms mean 'to separate' or 'discern', 'cut off', 'split' or 'refine'. Users view knowledge as a tool that helps to identify distinctions, yield to interpretation, can be discovered, be separated and analyzed, and can qualify and helps to achieve perfection. In Western Europe, from the middle ages to the Enlightenment, 'scientia' referred to a knowledge that was recorded systematically.

In other words, these notions and definitions of knowledge emphasize the maintenance of accurate knowledge – 'contingent certainty.' This definition emphasizes the need to overcome 'uncertainty' through the tools of decision-making and problem solving.

2.1 - Scientific Method is Evident in Law and Economics

There is a history of methods that had been used for increasing reliable knowledge. Upon closer examination, we can see that the scientific method is not unique. Law and accounting/economics/finance exhibit similar methods, processes and an evident knowledgebase. Similarly, the 'uniformitarian' ('algebraic') vs. legal historical foundational dichotomies are evident here also.

Figure 4 - Comparison: Scientific, Legal and Economic Methods

*1 – Historical Law-based – Geometric Natural Law; Legal Formalism
*2 – Empirical/Positivist, Rational, empirical, reductionist, axiomatic, postulates, social relativism.

Knowledge base: *'Conditional Certainty'*	Investigation of 'Uncertainty'
SCIENTIFIC	
*1 Geometric Natural Law; 'Laws of Nature' that are based on verifiable historical data.	*1 Decision-making and problem solving to help address and resolve uncertainty and bring resolved condition back to 'conditional certainty.'
LEGAL	
*1 History of authoritative documentation, precedents, and common law – quoted in the form of assertions, statements with citations. Known as Legal Formalism provides an executive level approach to Law. *2 This represents: positivism, legal realism, social empiricism and relativism in local interpretations. This is rule-based-supervisory approach to Law.	*1 Use to ensure clarity, provide a formal relational approach to describe legal evidence and aimed results. Legal analysis – is predictive, outcomes-based approach, which considers +/- outcomes and related consequent action. *2 This includes persuasive analysis – used for motions and briefs. Contingencies considered [Hypothesis, alternatives and weighing and justified (tests) judgments and verdicts.]
ECONOMIC	
*1 Normative economics (what 'ought to be') – planning for R&D based technology investments into innovative projects into infrastructural and 'manufactures' (Alexander Hamilton) projects in terms of recommending policy through representative government. *2 Positivist economics (what 'is') – focuses on economic phenomena and facts that are considered in terms of cause-effect, and is used to test economic theories. Considered being persuasive rather than descriptive economics – see post-/new-/Keynesian macroeconomics, monetarist, supply-side economics and others.	*1 Economic data is researched, identified, documented, quantified, categorized and is subject to all the investigative, strategic, policy, procedures, rules and regulations and other plans evident in the 3DMM (see below). Similarly data, conditions, events, rates are investigated forecasted and tested for reliability. These are typical processes evident in the scientific method – designed to resolve 'conditional uncertainty.' *2 This economics equated with those of Adam Smith and Malthusian options.

The scientific, legal and economic can either be erected upon uniformitarian principles or upon historical legal records. These link to 3D Management plans, Mental Models, 5 levels of Change Management, Decision-Making and Problem Solving methods for resolving uncertainty and returning to certainty. Similarly, management stylistic issues affect and guide these 'sciences': ethics (S1) and morals of a culture (S3) - methods, practices, hierarchy of values, in spite of what relativist modernist and moralist conclude (see Figure 5).

2.2 – The 3D Management Model (3DMM) Explained

All complex systems reflect management plans and infrastructures. The living system, engineering structure, scientific knowledgebase which reflect a 3D Management Model also become a knowledgebase where workable, measurable data, information, knowledge reflect conditional certainty.

The 3DMM (see Figure 5) reflects three infrastructural layers where management processes convert information inputs into output quality products. The infrastructures are management design, operations and style. Each of the three management layers contains nine relational management plans for 27-networked plans. Each layer has three rows: executive, supervisory and functional. Also contains three columns: directional, processive and data/information-base.

The multi-billion dollar engineering project management practices ensure the successful design, operation and maintenance of this 3DMM. This process involves multiple functions: project conception, planning, best practices, cost accounting, controls, processes, best world practices, quality, and implementation. These lead to meeting specific goals (short term) and objectives (long term, rewards). The product provides evidence of purpose, ability and capacity to meet specific objectives through concrete strategies and style. Two basic incentives drive such engineering project management: meeting specific objectives (deliverables) that result in the best short and long-term rewards (customer satisfaction). This streamlined 3D management process designed for efficiency and effectiveness, allowing it to meet performance standards, reduced costs, risk and timelines; ensure contract compliance; goals and objectives. Thus, such 3DMM becomes a source of verifiable, accurate and certain knowledge – ultimate goals of science.

These 3D management components are evident in all pre-set requirements in all living organisms and the management of knowledge. Knowledge management requires an approach that is just as rigorous in every aspect and part of its management exercise. Identified uncertainty within this Knowledge base kicks off the decision-making and problem-solving method (DMPS), which designed to convert uncertainty back to certainty. The DMPS is not a simple list of steps. The method instead includes at least five considerations:

a) Complete 3DMM - certainty warehousing (Figure 5)

b) The Scientific Method is the tool designed to resolve 'uncertainty' (Figures 9 – 13)

c) This 'uncertainty' can be inside or outside the 3DMM

d) Having examined the target area(s), this process searches history, standards, codes and the remaining stages of the 8-Point DMPS as it resolves the challenge (Figure 13).

e) Identifies the _hypothesis_ (alternatives - #4 Figure 10) and the _theory_ (quality guarantee - #6 Figure 10) within the 'uncertainty' realm that needs to be tested for certainty and quality guarantee (DMPS).

Conditional certainty can be planned for and be assured from the beginning – i.e., at the time when management design is being established. The definition of terms and concepts reflect processive and networked conditions. One avoids arbitrary definitions. Definitions must be set within the 3DMM related layers, 9 rows, 27 plans, etc.

Note that each of the plans has an address indicator. For example, the 9 plans on the *Management Design* level of the 3DMM have the 'D' indicator followed by an ordinal digit. D1, D2 and D3 form the Executive row. D4, D5 and D6 continue the flow on the Supervisory row, while D7, D8 and D9 continue on the Functional row. The same management plans pattern is evident on the *Management Operation* level – where the indicator is 'O;' and on the *Management Style* level – indicator 'S.'

In this configuration we find that D5 (Procedure); O5 (Controlling) and S5 (Attitude) are the only plans that link to all plans on their management level, while O5 is the only plan that links with all the 27 plans in the 3DMM.

Besides the Executive, Supervisory and Functional rows on each level (Functional, Operational and Design, each level is also:

Directed by columns S1, O1, and D1, as well as, S4, O4, D4, and by S7, O7 and D7; Processed by columns S2, O2 and D2, as well as, S5, O5, D5, and by S8, O8 and O8; Info based by columns S3, O3, and D3, as well as, S6, O6, D6, and by S9, O9 and O9.

It is clear that each of these plans is interlinked (networked) logically at their levels, e.g., D1, D2…D9, but also correspondingly with their counterparts at other levels O1, O2… and D1, D2… For example, a project manager will first, on Management Design, identify the 'Purpose' of the project (Executive, Directive), which when established, will affect (direct) not only D2 – Objectives, and in turn D2 – Strategies, but also foresee its impact on D4 (Policy), D5 (Procedures) and primarily D9 – (Productivity), for which it has been set in action to accomplish.

Figure 5 - The 3D Management Model (3DMM)

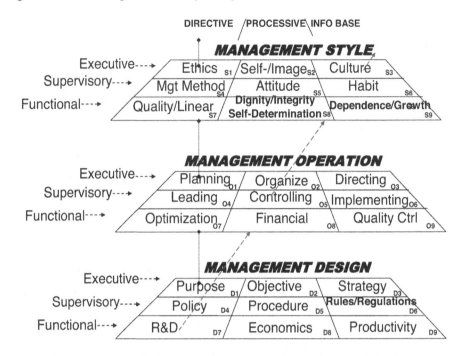

Table 1 provides a secondary means of explanation of the 3DMM infrastructures:

Table 1 - 3DMM analogy to Car-Driver-Manner

Level	Level	Analogy: car, driver, objectives
3	Management Style	Manner of driving – derived from awareness, ethics, attitude
2	Management Operations	Operational & maintenance instructions, effective use, objectives attainment with the most effective & safety use
1	Management Design	Product - vehicle manufacturing, standards, codes, production, reliability, customer satisfaction

The project manager will seek to 'automate' the nine *Management Design* plans in order to *Management Operate*, i.e., gets the project moving. In other words, while driving the vehicle - Design (D1-D9), the project manager will also anticipate and drive within *Management Operation* and *Style* levels plans. As an example, in the first book (*ECID*) contains an analysis of Leadership (O4), but in this current summary, the leader (O4) ensures and 'automates' all *Management Design* plans within the *Management Style* scope. In other words, the highly skilled and certified *(Management Design)* driver (leader), will perform *Management Operations* (plan, resource and operate the vehicle), within *Management Style* parameters and quality (ethics, culture, attitude and other plans). The driver's *Style* can be recognized by, among other things, by 'non-verbal' communication

- the way the tires are worn, unscheduled maintenance, repairs frequency, and other indicators. In the animal kingdom, all living systems reflect the 3D management model - the intricate infrastructures with their 27 plans and networked processes. Here, most management plans are pre-programmed 'automated' for specific networked functions and response to the environment. *Management Style* is 'wired' to ensure the living creatures' effectiveness in its environment; its negotiation parameters and resource selection. Creatures exhibit a purpose, objectives, strategies networked operationally with the automated functions. On the other hand, mankind's 3DMM's *Management Style* is programmed to consider more options. Should the network be compromised, man is able to select options, redefine and qualities improve many of the executive and management *Style* plans.

Both books focus on one of the three management infrastructures – Management Design (Table 2). Although the books provide examples as they apply to other infrastructures, the focus allows one to understand the basic infrastructure with its executive, supervisory and functional plans. This Management Design expands to provide evidence of its use at least three to four thousand years ago. Table 3, for example, shows the same structure and functions in the Tabernacle in the Wilderness, which prototypes the 3DMM, knowledgebase and features of psychology.

To better examine the *management design,* a parallel comparison below provides two examinations: a) 'business and project applications' and b) *'other applications.'* The first reflects definitions and applications as exercised within corporate management levels and plans; while the 'other applications' explore similar definitions within other fields: philosophy, religion, biological science – which reflect information management infrastructures, plans and networks.

The *executive level* details purpose, objectives and strategies (D1, D2 and D3 correspondingly). Each is these reflect more detailed sub-plan details, for example the purpose (D1) networks: law, identity and scope - each addressed in the content of 'business and project applications', as well as, 'other applications.'

PURPOSE (D1) - *Business and Project Application*

As summarized above, Purpose (D1) includes three components: law, identity and productivity scope and is set within a relational multi-dimensional environment. Purpose, with its networked sub-plans sets the groundwork for all levels, plans and activities.

- Law – is not a supervisory level 'rule'. In business, for example, Executive Law (D1) relates to contract that is enforceable by State, Federal, and International law, as well as physical/natural laws (e.g., 'acts of God'). *Supervisory rules* are derived from the Executive plans (D1-D3), policy (D4) and procedures (D5). Supervisory rules, together with policy (D4) and procedures (D5) support the implementation of the Functional plans (D7 thru D9) that are designed to lead to the quality product (D9) – thus fulfilling the Purpose (D1).

Table 2 - Management Design Level within 3DMM

Row	Directive	Processive	Info-base
Executive plans	**PURPOSE**	**OBJECTIVE**	**STRATEGY**
	Law	Priorities	Resource
	Identity	Standards	Forecasting
	Scope	Authority	Performance
	D1	D2	Plan (Budget, Gantt) D3
Supervisory Administration Operation Maintenance Safety/Security Emergency	**POLICY**	**PROCEDURES**	**RULES / REGULAT.**
	Guidelines to decision-making	Best/Economic practices	Do's and Don'ts Safety
		Processes	Security
		Flowcharts	
	D4	Software D5	D6
Functional plans	**POLICY**	**PROCEDURES**	**RULES / REGULAT.**
	Change	Accounting	Quality Assurance
	Improvement	Audit	Quality Control
	Innovation	Risk analysis	Output
	Invention		Waste
	Development		Delays
	Training D7	D8	D9

Purpose (D1) is in the Directive column. Purpose is linked directly to other Directive plans at the Stylistic Management's level - ethics (S1)and works to helps ensure successful Operational Planning (O1).

Executive laws imply contracts, agreements and/or covenants, which contain the terms, conditions, definitions, language, interpretation, jurisdiction, performance, judgments. Laws relate to and provide directive action to other executive plans: objectives (D2) and strategy (D3). Laws designed to lead to productivity (D9).

• Identity – The 'who' or 'what' that legally is within, outside in the environment, and contingent to the legal organizational process. Contracts identify who is within or without bounds, has security clearance, qualifications, and management responsibilities. Identity 'borders' define degrees of employee/management participation and contribution in an organization. Identify those who develop and contribute to the objective (D2). Defined Identities reflect specific policies and procedures to help reduce risk, and increase certainty.

• The productivity scope –identifies key processes that lead towards the fulfillment of the purpose (D1) – i.e., leading through economics, quality, performance, plans and infrastructures to the output and product (D9).

Both identify and productivity scope integrate with aw (contract) to establish the purpose (D1). These three define the purpose.

Purpose (D1) defines the WH-questions (who, what, where, when, how much/many and how) – that lead to the legal product in a legal & lawful (guaranteed) transaction within internal and external environments. Purpose (D1) directs the formulation and definition of the next executive plan: objective (D2) (i.e., priorities, standards and authority).

Curiously, the majorities of management textbooks hardly ever mention or mis-define the Purpose. Instead, the majority of these sources begin with the design of objectives (D2). Purpose does appear in project management texts, where the purpose appears in the form of a contract charter or the project definition.

How does the purpose (D1) reflect itself in non-industrial applications?

PURPOSE: Other Applications

Since Purpose (D1) reflects and helps define law, identity, and scope, this prime Executive row, when properly defined, will in turn help direct Objectives (D2) and Strategies (D3) and eventually all other linked plans – e.g., the Supervisory plans: Policy (D4) Procedures (D5), Rules & Regulations (D6), as well as, provide support for Operation plans and reflect Style Management.

• Law – The philosophy, ideology and religion describe two sources of law - executive and supervisory. These correspond to the geometric and algebraic worldviews.

The algebraic approach to law begins from a 'point,' which reflects a relative and subjective view (see also accompanying text to Figures 19 and 23). In management terms, Policy (D4) originates the point. The intellectual exercise (policy) using its directive initiatives postulates, creates an axiom from which a relative reality is reasoned out. This process includes the creation and interplay of algebraic variables (e.g., $a^2 \times b^2 = c^2$). Here we have axioms, theorems and postulates. The variables can represent anything. These variables may stand for materialistic, religious, mystical, economic and other designations. It is important to note that in this supervisory level process, the starting point is not the executive purpose (D1). Then, policy (D4) substitutes the purpose (D1) but uses the executive strategy (D3) to formulate its objectives (D2). This is a reverse action. This forms the algebraic value, prioritized filters. As mentioned above, this allows the filter to represent any variable – materialistic, mystical, or pantheistic. Having prioritized (Objective, D2) the resources (Strategy D3), everything then aligns with the improvised Policy plan (D4). Now the Policy plan, which has, like the Purpose (D1), a Directive (column) effect, will attempt to 'justify' (legalize) the Supervisory plans: procedures (D5), rules and regulations (D6). This in turn ensures the Functional plans (D7, D8, and D9).

This approach is similar to having the tail wagging the dog principle. In the philosophical/religious terms, this translates into rationalistic, empirical, idealistic, reductionist and pantheistic views. Similarly, mythological, cult, pan-/poly-theistic and the various versions of most monotheisms are essentially built on this algebraic relativistic structure because these do not have a legally based Purposive (D1) determinations (e.g., geometric natural law). This makes them become subjective and relativistic. This is why, when philosophers, materialists and many theologians debate on issues of God, while basing their reasoning on the algebraic approach, their definitions of divinity will inevitably be pan-/polytheistic. Such divinities will appear in the form of Prime Causalities, mythological or teleological constructs. This factor helps explain why algebraic-based worldviews have historically fit comfortably within the (Roman) Pantheon of cults.

A typical application of this relativistic principle in ethics is seen in the approval by the Human Fertilization and Embryology Authority (HFEA) to justify researchers at the University of Warwick (UK) to proceed with the human-pig embryos hybridization. HFEA will 'issue a license following stringent checks [that] demonstrate that it is considered both necessary and ethical.'(Highfield, R. 'Human-pig hybrid embryos given go ahead,' posted on Telegraph.co.uk July 1, 2008.) This was a decision based on axiomatic premises. There is no consideration given to the justification for such a landmark decision. A simple committee determined that it was ethical and it became ethical by virtue of a Policy statement. This shows that such axiomatic ethics drawn upon an authoritarian/totalitarian paradigm. By extension, historical evidence demonstrates that such foundations for decision-making have led to allow the institution of concentration camps, genocide and the legalization of cannibalism.

Prior to Vatican II, the Roman Catholic Magisterium had at least taken into consideration thousands of years of precedents. These precedents originally emanated from divine law, the scriptures and through holy traditions. However, in the above cited relativist authoritarian committee vote approach, it is the committee that sets the precedent without any legal requirement. Sociology drives this decision-making structure. Empiricist rules and relativist variability are the rule.

In contrast to this algebraic approach, the geometric view of Law begins from what the temporal human mind (psyche) can interpret in terms of geometric concepts. Here, reality begins not with a 'period' (algebraic approach – see Figure 19) but with a circle (see Figure 20). This geometric process also includes the conical multi-infrastructural qualitative system (see Figure 21) and the five convex regular polyhedral solids (tetrahedron, cube, octahedron, icosahedrons and dodecahedron). Plato described these solids -'Timaeus' 360 B.C., but also appear described on stone monuments, in Scotland – 1000 years earlier[ix]

These geometric features are built within all of creation at the physical, sub-atomic, solar system (see Johannes Kepler's, 'Mysterium Cosmograhicum' Tubigen, 1598) to super-galactic clusters levels. This geometric feature is also evident in the musical scale and physiological proportions in animals and plants.

The original Christian Bible is unique in that it is a legal historical document consistently reflecting geometric natural law. This is evident from the first chapter of Genesis (see Figures 24 and 26 for detail). The first Genesis chapter identifies its executive purpose, law, identities, and scope. In the first chapter, the author identifies Himself as the Creator – the prime covenant designer (Chapters 1:1 to 2:3). The creative process of ex nihilo (out of nothing) is easier to understand within the geometric concept. Biblical communication and conversations reflect a legal form. Here we find documented events, precedents, laws, blessings and curses – see also Deuteronomy 27: 15ff; 28:1 for detail. Later there are the secondary covenants with patriarchs and kings that are the underlying themes for understanding the Old and New Testaments. It is worth noting that today when geometric natural law is not consciously applied, then Christians tend to provide substitutes – e.g., inerrancy and literal/-ist interpretations of the Bible.

The identity of the second party to the Covenant is retained as the Chief Executive Officer of all Creation (Genesis 2:19). Many writers have seemingly found contradictions and paradoxes in Genesis chapters 1 and 2. However, these issues are resolved when seen within covenant, legal context of the Christian Bible. God himself wrote Chapter 1 – he signs His name as the 'supreme God' (Gen 1:1-1-2:3). In Genesis 2:4, a second person describes this God as being the 'Self-Existent supreme God' (Eternal Lord God). This second view is typical of how the human mind would perceive God within geometric natural law – the 'circle': self-existent supreme god - the Prime Covenant giver. Thereafter, descendants of Adam – the patriarchs track the Covenant in legal documentary format from generation to generation. These patriarchs are clearly identified as members of the second party to the divine Covenant: Gen 2:7 – man (adam); Gen. 5:1 Seth; Gen 5:7 Enos; Gen 5:10 Kenan; Gen 5:13 Mahalalel, to Gen 5:30 Noah; and post-Flood patriarchs Gen 10:1 Shem, Gen 11:12 Arpachshad, Gen 11:13 Shelah, to Gen 11:27 Abram, and beyond.

All these legal and historical records have been accumulating from before and after the Flood. Prince Moses of Egypt finally summarized this veritable legal and historical library, using the Egyptian script, in the document now known as the Book of Genesis. With this perspective, the reader should view the Book of Genesis in terms of a reflected original veritable legal library. This document represents the Knowledgebase - the 3D Management Model set within geometric natural law.

We have a record of the legal family descendants who maintained the covenant of righteousness, maintained the legal and historical documentation, laws and judgments throughout history. Here also we have a record of three singularities that had global and universal implications.

The first singularity is the creation of the original perfect Creation in seven days. This includes the seventh rest day. The creation week may be considered as a single overarching singularity, or in terms of seven separate sub-singularities because each were marked by seven separate volitional and unique events, each reflecting distinct infrastructures, resource loading and distinct geometric solids (see Figure 26), as well as quality control evaluations – 'it was good'. The same Prime Party of the Covenant has initiated these creative events and unique legal, letter-number correspondence style, thus musical – as a marker within creation by the Creator. None of these elements and features are ever mentioned by uniformitarian scholars. At the same time, it has become evident that uniformitarians have formed numerous schools of linguistic and historical research and interpretation to help hide these features from the 1st chapter of the book written by God.

The second singularity occurs when the perfect creation's energy density drops one cycle down the cone's infrastructures (see Figure 22). This 'energy-density' leads to a condition of 'energy starvation' throughout the original creation. It is where 'death' enters the timeline. This condition changed and reduced cycles, velocities and re-defined the second law of thermodynamics. This operational interdependence established secondary relational, networked and infrastructural relations. This event occurred on the 11th day on the Creation calendar when Creation's CEO (Adam) made a critical imperfect decision.

The third singularity appears in less than 2000 years after the first and second singularities. The third singularity begins with the initial global hydro-tectonic catastrophe lasting 12 months and 10 day and is followed by secondary environmental adjustments that are mostly evident in the current geologic, paleontological conditions, structures and morphologies. The initial hydro-tectonic global catastrophe included continental separation and mountain building. The secondary catastrophic self-adjusting phase continued for several hundreds of years that included events of mega volcanisms, climatic change – a *circa* 500-year ice age that included torrential rains along the tropical zone. During this time, ocean seas rose between 500 feet at the equator and 100 ft (north/south of the equator), post ice-age desertification – e.g., Sahara, Middle and Far East. The Giza Pyramid and the Sphinx originally stood in lush vegetation with lakes nearby. After the Ice Age, these structures stood in the middle of a desert. The Bible records the initial catastrophic events about 1656 years after the first day of Creation (Gen 1:1). The Flood events continued for 12 months and 10 days (Genesis 7:11 to Genesis 8:13-14). The secondary - self-adjusting phase - continued thereafter primarily for hundreds of years, and continued in diminishing severity until today. The existing polar icepacks and those in the high mountain ranges is all that remains of the 2000 BC to 1500 BC ice age, and these are now beginning to melt also.

A detailed presentation of geometric natural law is necessary because it sets the parameters for understanding all of the following evidence that the Creation model has to offer.

The Intelligent Design group does not have an official origins theory. Its laboratory approach focuses on laboratory empirical data. From this, they interpret data and empirical evidence. From this assemblage and categorization of data, they may describe features of purpose. Here they describe physical law that is implied at the sub-chromosome information levels, biological irreducibility and harmonizing constants of the universe. They will hypothesize about the concept and participation of an Intelligent Designer. They will also speculate teleologically about the scope and purpose of all of this fine tuned universe and sophisticated creation.

This section distinguished, scoped and provided application examples of Purposeful (D1) law. Executive Law (D1) is clearly distinguished from Supervisory Rules (D6). Some people may talk of rules as laws, but in this case, such laws begin with a small 'l'. Executive Law represents geometric natural law; while supervisory law, unless aligned with Executive Law, will express algebraic axioms.

• *Identity* in the 3DMM Purpose (D1) refers to four areas of an organization: what is inside, outside, support (contingent, supplier), or a customer (produce receiver) identity.

The identity's job description establishes a legally functional relationship with individuals, groups, organizations, institutions, etc. An identity functions within geometric natural law (GNL) recognizes the 3DMM dimensions. Such a user aligns the multi-dimensional links from the Purpose (D1) to all other plans on every infrastructure. The user recognizes man's executive faculties and respects mankind's 'divine spark'. The GNL user's approach recognizes the knowledgebase of best practices, and standards. Approved identity accesses discoverable evidence and management stylistic quality (S1 – S9), the supportive competencies, skills, creative and productive benchmarks are part of the identities.

In the Christian Bible, the Creator is the Prime Party of the Covenant, while Adam is the second party to the covenant. Adam with a title of CEO of Creation and even though he missed on a key decision-making issue, his righteous descendants have carried that covenantal title from one generation to the next. The terms and conditions that derive from the Purpose, as they implemented and derived from all infrastructural levels (Design, Operations, and Style), direct the righteous in all aspects.

There are five types of 'outsiders,' who are <u>not</u> part of the covenant/contract. These are the passive or potential customers or partners; external or competitor's intelligence gatherers; competitors and antagonists. In the first chapter of Genesis, we find two types of 'outsiders.' One group that is covenantally unqualified - the living creatures that cannot become Adam's helpmate (Gen 2:15-20). The other is Eve who created from part of Adam's body and does qualify under the covenant; however, Eve's breaching of a portion of the covenant had made her an anti-type - an antagonist with regard to the righteousness of the covenant.

Eve has become an antagonist by having prioritized supervisory rather than executive thinking. Doubt – an axiomatic position, is begun with a hypothetical question - 'what if the purported A is not really A, but is really B?' It is supervisory thinking because such 'purely' rational decision-making – is a Policy (D4) function through a substitute plan within the directive column. She re-prioritized the Executive Strategies (D3) and Objectives (D2) to now filter (priorities, standards, value hierarchy) Policy based decision-making and problem-solving (DMPS) (fifth step) (see details below). This reprioritized filter bypasses the Purpose (law, identity and scope) and provides a substitute – an empirically based (touch) rationalization (wisdom) that satisfies the subjective (food). In this context, all 3DMM-networked plans are affected. The antagonist: 1) challenges the covenantal relationship through an alternate attitude and self focused subjective ethics plan; 2) maximizes risk (death) and vulnerability (threat); 3) reprioritizes security (hides) by inflating self-image (eyes were opened); 4) reformulates culture, the level upon which, and within what parameters (levels) communication is conducted, etc.

The *outsider's identity* reflects an alternate worldview – a 3DMM that may be subjective and relativistic. For example, when God asks what had changed Adam's original view (perfection). Adam presented his subjective causalities – 'accusing' the faulty woman (sub-value) that God gave him, and the Woman faulted the serpent. Both showed that they had rejected the perfect relationship covenant, where everyone loved (10 commandments) and cared for the other, and both loved God.

Clearly, Adam and Eve had amended the original covenant. They derailed and re-prioritized the original executive, supervisory and functional plans. They produced an alternate product to satisfy another customer.

Mother Eve opened an alternate way of thinking based on rationalism (policy) (D4). As mentioned above this Policy plan is located in the directive column, and is on the supervisory row. This approach bypasses the Purpose (D1), which is also in the directive column at the executive row. Eve prototyped the 16th century rationalist, empiricist, reductionist and 19th-20th century's modernist methods. These used the algebraic approach to promote relativistic, subjective and quantitative (economic) evidence. Inevitably such proponents introduce and recognize the application of the rule of strength, cunning and a ruler-servant, master-slave (masses, labor, democratic, collectivist) systems. This appears in the establishment of a Babylonian worldview.

We should not forget the original Edenic message and pattern of the 3DMM. The Bible contains the 3DMM's design management plans in the form of the Tabernacle in the Wilderness (Exodus 22 ff) and Table 3 describes the structural parallels between the Tabernacle and the Management Design plans of the 3DMM.

Table 3 is a combination of the Management Design infrastructure within the 3DMM (see Table 2) and the Biblical Tabernacle framework (Table 3). There is a correspondence among the infrastructures: executive function (spirit); supervisory functions ('psyche'/soul psychological); and functional (physiological). On Table 3 the Tabernacle, plans are italicized.

Table 3 - Management Design in the Wilderness Tabernacle

Row	Directive	Processive	Info-base
	PURPOSE	**OBJECTIVE**	**STRATEGY**
Executive plans	Law	Priorities	Resource
	Identity	Standards	Forecasting
	Scope	Authority	Performance
	D1	D2	Plan (Budget, Gantt) D3
Holy of Holies *Holy Spirit*	Law *(10 Commandments)*	*Aaron's Rod that Budded* Priesthood (authority)	*Dish of Manna; Holy Scriptures Prophecy (Forecasting)*
	POLICY	**PROCEDURES**	**RULES / REGULAT.**
Supervisory Administration Operation Maintenance Safety/Security Emergency	Guidelines to decision-making D4	Best/Economic practices Processes Flowcharts Software D5	Do's and Don'ts Safety Security D6
Inner Court *Soul/Psyche*	*Seven Candlesticks Intellect/Reason/ Decision-making*	*Sacrificial Table Will (power)*	*Sewbead Emotion*
	POLICY	**PROCEDURES**	**RULES / REGULAT.**
Functional plans	Change Improvement Innovation Invention Development Training D7	Accounting Audit Risk analysis D8	Quality Assurance Quality Control Output Waste Delays D9
Outer Court *Physical functions*	*Search Area Sensory/scanning*	*Wash Laver Filter & Account*	*Sacrificial Table Productivity & Quality Control*

Briefly, Executive plans (D1 to D3) correspond to the Tabernacles' Holy of Holies, which represents the Spirit, where the Purpose (D1) (law, identity and scope) here are the 10 Commandments tablets. Where Objectives (D2) (authority, priorities and standards) correspond to priesthood and authority. Where Strategies (D3) (resources, forecasting, performance, plan) correspond to the pot of manna or spiritual food – the scriptures – i.e., knowledgebase. The Tabernacle supplements executive plans with vision = mercy seat (Ark of the Covenant).

How does this correspond to living creatures? Living creatures reflect Purposive adherence to natural biological laws, identities (e.g., syngameons, species, and taxa) and scopes. These, and supervisory (psychological) and functional plans are pre-programmed (instinct). Such pre-programming allows for minor physiological adaptation, within the syngameons scope, to challenging environments.

Algebraic reductionist/materialist proponents describe living entities in terms of mechanic-robotic-biological or processive patterns that change due to environmental challenges in linear statistical relative progressive increments (mutations and survival of the fittest). Upon closer empirical examination, these are at best descriptions of supervisory-level concepts and not descriptions of Executive-level infrastructural levels. The large-scale tools and concepts (mechanic, robotic, biological) leave out the information-software-infrastructural and management plans networks that are required for living entities. Neither does this account for the five-management change levels (see Figure 16). Evolutionary models must account for causes that change management executive processes – **purpose** (law, identity and scope), **objectives** (priority, standards and codes) and **strategies** (planned resource forecasting and performance). These management plans network and direct the supervisory and functional infrastructural operations that lead to specific productivity. These, which are formulated at the purpose (D1) productivity scope, lead directly through networked infrastructures to several plans to production (D9).

• Productivity scope is the third of the three executive purposive operations. This scope is the coded process that formulates the direct path via the supervisory infrastructure to quality-based productivity [D9].

The *algebraic (uniformitarian) method* explains all life forms in terms of reductionist methods - mechanical, biological or processive mental model. Man's mind, it is alleged, can only be understood in terms of behaviorist action-reaction processes. That is, only through supervisory plans - policy (D4) – conditioned decision-making and problem solving; procedures (D5) – enhanced conditioned practices; rules & regulations (D6) – emotional conditioning. In contrast, the GNL-based management includes the executive plans (D1 through D3). Living beings have a determined network that specifies established processes from the executive to supervisory infrastructural processes. These may be pre-programmed (instinct) (habits – rules) and lead to secondary functional processes, interaction with the external world, search & development, economic inputs, processing and outputs through health, security, safety resulting in productivity.

The larger volume - *ECID* provides a more detailed explanation; this summary's objective is to focus on Figures and Tables that are tools for the Scientific Score Card.

Here is an example of the application of Ethics (S1) within *Management Style*. Figure 6 compares how various debaters and users interpret Ethics (S1).

Figure 6 - Types of Ethic - Definition of Man

Evolutionary	
View of man	Materialistic relative changes lead by chance (statistical progression) within a challenging environment, from material to single living cell-multiple cells, to primates and man.
Definition	Man is an upgraded animal with specific biological and determined behavioral needs, subjective relative wants, and social conditioning (e.g., behaviorism)
Management level & Plan	The reductionist view removes 3DMM area: a) Management Style; b) Executive plans/rows. Instead, focus placed on Habits (S6), implementation (O6), Rules & Regulations (D6), Procedures (D5).
Commercial – Mercantile	
View of man	Emphasis on consumer, producer, and market plan (needs, wants, satisfaction and delight); social economic demographics (lower, middle, upper class, temperaments).
Definition	Man is an upgraded animal with specific biological and determined behavioral needs, subjective relative wants, and social conditioning (e.g., behaviorism)
Management level & Plan	The reductionist view removes 3DMM area: a) Management Style; b) Executive plans/rows. Instead, focus placed on Habits (S6), implementation (O6), Rules & Regulations (D6), Procedures (D5).
Marxist (social classes) & Nazi (racial distinctions & qualities)	
View of man	Views suggest a) irreconcilable distinctions among proletariat, bourgeoisie and capitalists. In this environment, we find key values – existence of a *Dictatorship of the Proletariat* and *Atheism* at the exclusion of all other notions; b) similarly, the Nazi value is identified in the higher race vs. lower races; *Dictatorship of the 'folk'* and *Atheism*.
Definition	Proponents of these views divide society by any possible variables: class, race, age (fetus/old). Three categories exist in: 1) producers, 2) producers exploiters, and 3) those who manage producers' interests and production. Producers form a collective and own everything in common. Evolutionary views – society evolves from 2) to 1) via 3). The other view categorizes society into 3 types of producers: 1) highest productive groups; 2) lower less productive and purely consumer /parasitical groups, and 3) those who manage types 1) and 2). *Evolution* filters 2) to 1) via 3).
Management level & Plan	Both Marxist and Racist views are located in three areas on 3DMM: 1) *Info-base column* (D, O, #3, 6, and 9); 2) *Management Style – Culture*: where they see: collectivism or power of the will (S3); 3) *Habit* – working class or productive and ingenuity (S6), and Dependency (S9). State authoritarianism with absolute enforcement manage this evolution.
***Supernatural* - Christian (not cultist, which falls under any of the above)**	
View of man	Man made in the 'image of God.' Man reflects the Creator. They have respect for the value of family, brothers and sisters – broth's keeper. Authority and members recognize geometric natural law, 10 commandments.
Definition	When this view is not cultist, the existence and communication of the supernatural entity must provide: 1) Geometric natural law at the foundations for Creation; 2) Historical contractual documentation with man, 3) History case histories, commandments, applications, statutes, precedents, judgments and directed purpose, i.e., Objective law.
Management level & Plan	Addresses all 27 Management Plans in the Management structure and must reflect this at the individual, social, organization and intra-national level.

The 3DMM identifies Ethics (S1) on the Executive row (S1-S3) and in the Directive (S1, S4, S7) column on the Stylistic infrastructure (S1 thru S9). However, the marketplace, organizational, social activities and the uniformitarians substitute this Stylistic Executive level - this 'ethic' is degraded to rules and regulations (D6) on the Design Supervisory infrastructure. Here are some concrete examples of pseudo-ethics.

Today we have *'normative ethics'* set on the Management Design Supervisory area (D4, D5 and D6). This normative ethic consists in articulating, implementing and having people to perform through authorized and approved policies (guidelines to thinking). These policies imply prescribed procedures enforced through rules and regulations (D6). It is here that right and wrong, good and evil are determined. Developers of this Supervisory rule-based approach are Aristotle (virtue ethics), Emanuel Kant (deontological normative theory) and John Stuart Mill (utilitarian, consequentiality).

This *etiquette* comes from situational conditions. A collection of observed human behavior helps formulate such situation ethic. It is the observations of actual choices made in practice by the hands-on users, buyers and voters. This observation of 'collective' behavior and views does not require expert or specialist advice. Observers of this collective behavior collect filter and then transmit this 'ethic' through story telling or are broadcast as simple 'common sense'. Since 1978, Judith Martin ('Miss Manners') has provided such filtered 'collective' observations as advice in the United Features Syndicate, (200+ newspapers worldwide). She answers her readers' questions on etiquette and writes short essays on problems of manners or politeness. She reflects the type of etiquette, situational and the applied rules-types approaches that are evident in areas like business, politics, medicine, environment, ecology, journalism, jurisprudence, engineering, genetics, abortion, feminism, sex, academia, tenure, education, psychology and economics. Each area seems to have its own ethic derived not from the 3DMM Executive levels but from operational and functional levels.

Meta-ethic is another Supervisory type of ethic (etiquette). This Supervisory *ethic* places value on: a) people's feelings (emotivism), b) their interests (prescriptivism) and c) belief systems (cultural or individual relativism). These three factors determine moral values. To explain these phenomena one-group suggests an evolutionary development of the human psychology. Another group suggests that the 'collective' debate among people leads to ethical standards. Another group points to some kind of objective independent truth that is intrinsically set in the world. Moreover, if so, then some theology and naturalistic philosophers suggest the discovery of ethical principles that may be measured for validity.

Some authorities suggest that *meta-ethic* be broken into 'non-realism' and 'realism' schools. In addition, others emphasize *self-esteem*, which is a close synonym to Self-Image (S2) on the Management Style level. Yet again, these also remain within the Supervisory Rules & Regulations (D6) scope. In other words, they are algebraic axioms and theorem-based determinations, and don't reflect the Executive level Ethics (S1) at the Style level, neither do they network within 3DMM with other management plans at executive and supervisory infrastructures.

In contrast, original Christianity has a clearly identifiable Purpose (D1) which is fed through two infrastructures (Style and Operations) based in a 3DMM specifically from Ethics (S1). This Purpose (D1):

a) Derives from geometric natural law
b) Aligns with Ethics (S1)
c) Where the 10 Commandments exit, they include the Covenant foundations that contain key components for personal, social and civilizations' self-perfection
d) Documents social and civilizations' standards expanded from c) and guarantees 'blessings or curses' (Deut 27-28) thus
e) Distinguish among those citizens who qualify for the Kingdom of God, and those for the Kingdom of Babylon
f) Maintain a knowledgebase of accurate information, knowledge and wisdom including continuous self-improvement and standardized self-perfection

The 3DMM summarizes the complexity of what goes into the workings of any complex management system (e.g., engineering project management, information, knowledge base). Such information systems involve databases and knowledge base systems, and here we can test other information and knowledge bases such as philosophical, political, ideological and religious systems. Furthermore, human language reflects an information and knowledgebase that reflect 3DMM group of management infrastructures: 1) phonetics/ lexicology grammar (Management Design); 2) syntax (Management Operations) and 3) stylistics (Management Style).

Here's another analogy of the 3DMM which the language literary levels reflect:

Figure 7 - Comparison: 3DMM with Language/Literary Levels

3DMM – Language Literature Levels – Observed Parallels and Similarities
MANAGEMENT DESIGN: D1 – D9 – Grammar
The composition of language and literature begin with the language elements (parts of speech): verbs, nouns, prepositions, etc. When arranged meaningfully they demonstrate a *purpose* (laws, identity, scope); *objectives* (standards, authority), *strategy* (forecasted utilization), in accordance with specific cultural conventions (guidelines - *policies*), arrangements and order (*procedures*) and *rules*. These are edited (*R&D*), for coherence and clarity (*economics*); and communication, meaning and audience/customer appreciated results (*productivity*). Inevitably, a grammatical sentence reflects purpose and objectives (*standards, authority*).
MANAGEMENT OPERATIONS: O1 – O9 – Syntax
Each communication reflects specific objectives. The writer refines communication through planned, organized and directed use of expression alternatives: simple, complex, compound or any of these combinations that lead to clear objectives - desired results. Writers or speakers lead their audience by using controlled techniques and qualities that help implement objectives. Writers, speakers, and artists examine interconnection of/among ideas, concepts, coherence and relevancy, here expressed in each sentence, implied meaning and sub-/structures to bring out value (*optimization*). Focus on the *economy* of word and image lead towards precise, durable and

clear ideas and action. These in turn lead to the attainment of proper audience desired effect and results (*quality assurance and control*)

MANAGEMENT STYLE: S1 – S9 and Stylistics
Authors refine textual content & context with effective semantic, synonymic, symbolic, typological application. With this, they aim to achieve effective, representative, reflective, uniform communication, and a perception of and possible habit change (ethics, image, and culture). Over time, writers have considered six stylistic levels of communications. They considered from the lowest - 'vulgar' with a maximum 500 word vocabulary that lasts 5 years, to other levels - common, technical, literary, philosophical and the highest - poetic level with its 200,000 word vocabulary and 250 year longevity. Stylistics involves establishing the author's writing skills, methods, attitude and themes. As a manager of his craft, the author or artist will identify qualitative contribution, dignity, integrity, originality (self-determination) and productive expression of freedom or dependence. Style does not only focus on proper language usage (habit, dependence, culture) but uses all available communicative means (culture, attitude, methods, images, ethics, musicality) to ensure that the entire work (short story, article, novel, music, artwork) is in total harmony – (3DMM) managed – all 27 plans at the three infrastructural levels, i.e., having each part contributing to enhance the whole. The author will respect dignity, integrity and self-determination of all or some of the audience.

2.3 - Uncertainty Management: Supervisory Plans in 3DMM

Where the *Management Design* levels *Executive* row ensures '**conditional certainty**, it is the *Supervisory* row that help address '**uncertainty**' wherever it may appear, this section examines the means and tools for overcoming 'uncertainty' – Policy (D4) – i.e. guidelines to thinking; or more specifically guidelines to 'decision-making and problem solving'.

There are dozens of decision-making and problem solving techniques. Many formalized techniques appear as *procedures* (D5) - computerized, automated and neural networks; or as '*rules and regulations* (D6). All of these procedures and rules & regulations are programming techniques that reflect proven 'laboratory' methods for reducing 'uncertainty.' They are useful in their second and third stage of ensuring 'conditional certainty.' Procedures may be rigorous but they are not laws. They will vary with environmental conditions and standards/codes of produced products. A change in product, system, process will require alternate adjusted procedures (process, software). Rules & Regulations, however, are pre-set for ensuring safe and secure operations.

Users and managers must 'trouble shoot' unknown phenomena. They encounter new variables, alternatives or must overcome barriers; in such cases, they will use a wider field of the decision-making and problem solving method. In the initial stages, an organization or system provides the Executive parameters – the Purpose (D1), Objectives (D2) and Strategies (D3) within which trouble shooters operate – these Executive parameters provide the 'guidelines to thinking.' These guidelines reflect the thoughts, standards and tests in a competitive environment. Doing any, type of thinking outside these Executive parameters will simply reflect thinking within the objectives and strategies used by the competition – thus exhibiting collusion, treason, etc.) However, should the Executive identify new opportunities within the scope of the Five Qualitative Levels of Change Management and create a new organization with a new Executive scope – Purpose, Objectives and Strategies, then this will legitimize a wider field of decision-making and problem solving methods.

The purpose for applying the decision-making and problem-solving method is the same as that for the Scientific Method - to find a solution (conditional certainty) when uncertain conditions exist.

At its foundation, decision-making and problem solving may be very simple. For example, we may compare the challenges that a creature – e.g., mouse must overcome with that of Christopher Columbus:

Example 1: A mouse has to go through a maze to get at the cheese

Example 2: Columbus must persuade sponsors, and then must discover a route to India (America).

Both the mouse and Columbus consider and implement the following actions:

- A situation demands resolution
- Time pressure
- Lack of complete information
- Uncertainty suggests risk for any decision made
- Consequences expressed in terms of reward or punishment
- Existence of two or more alternatives (contingent) action
- Adjust for personal qualities – strengths, weaknesses, competencies

These considerations can provide opportunities, affect the decision making process, or affect the manner in which problems are resolved. Since 'uncertainty' emerged, the qualifications of the decision-maker become the first consideration – it is really the first point to consider in the 'uncertainty' factor – see Figure 8

Figure 8 - Personal Qualities Affecting Decision Making

Mental:
    • Executive faculties: purpose, objectives and strategies
    • Supervisory faculties: *Intelligence* – decision making and problem solving
                           Will – motivation, procedures, flowchart
                           Emotions – data, information, habits, patterns, rules
 Physical: Senses (scanning), memory, neural system, health
 Skill: Abilities, capabilities; competence and competencies, experience with 5
     levels of change

Critical investigative mental process is vulnerable to a myriad of traps. The book *ECID* examines dozens of these traps. For example, two professionals may apply the same logic, examine the same data, and yet may present radically different 'solutions'. Under these conditions, we may have thought that having enough Data + Logic = Solution. However the disagreement on the solution proves that Data + Logic = Alternatives.

The first part – 'Data' is neutral – it is not a 'fact' and not even information, since information is contextualized data. For example, the numerals 6, 12, 43 mean little unless we know that each separate numeral represent the day, month and year of a date – June 12, 1943. So 'Logic' functions differently when it is in contact with 'data,' 'information,' 'system,' or 'wisdom.' If logisticians agree on the scope, and several of them may agree on the means to a solution; this situation may still reflects a 'conditional certainty.' This is due to the existence of variances – 'alternatives':

$$\frac{Logic + Data}{Interpretation + \Pr iorities} = \textbf{Alternatives}$$

These are interpretations,

Evolution, creation, and intelligent design scientists may look at the same phenomenon (e.g., fossil in some sedimentary strata) but each may present different interpretations and priorities. This diversity of opinion means that instead of having achieved conditional certainty, only 'alternatives' are present. We may go a step further and consider that perhaps two of the three groups of scientists have actually reached the right solution (certainty), but since one did not agree with the solution or interpretation, these solutions remains at the 'alternative' level, and this is why the debate continues today.

The basic decision-making and problem-solving (DMPS) (scientific method) is designed to resolve 'uncertainty.' Steps 1, 2, 3, 4, 5 6, help achieve 'conditional certainty,' while the resolution of #7 provides extra qualitative guarantee – a step that may help achieve absolute certainty. However, both #6 and preferably #7 represent the knowledgebase of general conditional certainty in #1 in the DMPS. Note that trouble-shooters often miss on the hierarchy of terminology. For example, it has become common for some to ascertain an 'almost certainty' whereas #5 (Filter, Choice Making) has pre-conditioned all other terms 1, 2, 3, 4, 6, 7 thus rendering the DMPS or the scientific method to remain in 'uncertainty' become meaningless.

The basic *decision-making and problem-solving* (DMPS) process contains seven activities. It begins with *conditional certainty* (#1) or 3DMM – accurate knowledgebase. This 3DMM reflects, as mentioned above, the 27 plans laid out on three management infrastructures - Design, Operational and Stylistic These 27 plans are not necessarily perfect, they contain degrees of 'conditional certainty' but they are perfectly workable. However, the

Figure 9 - Basic Decision-Making & Problem-Solving (DMPS)

Decision						
Problem Solving						
Decision Making						
	Decision Analysis			Choice Making	Decision Taking	
1	2	3	4	5	6	7
Certainty	*Uncertainty*				*Conditional Certainty*	
Programmed	Psychological	Definition	Alternatives	Filter	Implem/ Test	Quality Guarantee
Mgt Design Operational style 3DMM	Mental: purpose objective strategy Physical Skills	Define challenge in terms of necessary outcomes. Tools Time Diagnosis	Consequence Opportunities Quantify Borders Consequence	Priority Weigh Validate	Budget Plan GANTT Standard Quality Ctrl	Reliability Low risk Purpose to go to 3DMM

environment and the internal 3DMM plans and infrastructures can catalog a range of certainties. Some of these may be workable, while others may range from being semi-effective to unworkable products, systems, designs, quantitative information. Yet they may be useful for reliable application, management and research. In this environment, *formal problem solving* may not be required – a conditional certainty guarantees that the system will work without failure within the parameters established. However, when something compromises certainty (internally or externally) uncertainty emerges and requires DMPS application to help resolve the challenge, and bring the situation back to conditional certainty. The debates: Evolution, Creation, I.D. and Hybrid Uniformitarian, is a contest in demonstrating which of these scientific models offers the best scientific interpretation, prediction of data, information and offer a reliable knowledge base. Here are two examples of addressing conditional certainty vs. 'uncertainty.' To ensure certainty:

Example 1: a transportation vehicle is designed to function in compliance with the highest engineering standards. It must meet external operational conditions and customer expectation. Prototype vehicles tests must pass manufactures' inspection and guarantee warrantees. At the same time, any vehicle malfunction introduces a condition that must be resolved through DMPS in order to return the vehicle back to its certainty status.

Example 2: The Christian Bible describes that God first created Time within Infinity (infinite circle) - thus reflecting the 'canvas' of geometric natural law. God created geometric solids, their infrastructures and content within designated days. The completed work reflected perfection and certainty 'it was good'. However, by creating man with free will – i.e., man having the capacity to exercise decision-making and problem solving, posed a risk. Man, being a finite creature (created within Time), with supervisory features (soul – intellect, will, emotions) that may or may not be aligned with executive functions could introduce imperfection (uncertain, un-lawful) features within perfect managed certainty. Adam did introduce uncertainty and had to solve a triple challenge: 1) imperfection affected all of creation, since Adam as the CEO of Creation affected his entire realm with imperfection. This imperfection brought about the second singularity – where all reality dropped one energy density level; 2) because of imperfection, Adam was not able to correct the problem – imperfection affected and downgraded not only nature but also him – his decision-making and problem-solving faculties. The DMPS tools were now limited to and have become the product of the second singularity not the first (Creation); 3) the new DMPS was 'grounded' to the economy of the imperfect, or conditional certainty. Adam found himself where he had to use degraded tools, software and lower technical competencies to repair, for example, a high tech anti-gravity aircraft. In another example, we may have scientists with an 'organic model of thinking' (see Figure 15) attempt to interpret information engineering and nano-robotic processes at sub-chromosome levels.

Figure 9 shows that 'uncertainty' (steps #2-#6) begins at the psychological stage (#2). This psychological state, unless properly addressed will lead to freezing the process on the 'alternatives' step (#4) level and will continue ensuring 'uncertainty' not certainty. At the psychological level, the assumptions, beliefs, attitudes, value, goals, objectives and needs / wants come into play. Figure 10 provides additional practical detail, with two examples for each stage:

Figure 10 - Normal Steps in DMPS

3	4	5	6	7
UNCERTAINTY			Conditional CERTAINTY	Quality/guarant. CERTAINTY
Definition	**Alternatives**	**Filtering**	**Test**	**Qty Assessment**
A gap/ problem/ 'uncertainty' appear. Objective is to convert 'uncertainty' into conditional certainty (#6). Therefore, 'define'(#3) the 'uncertainty' in terms of the desired solution (#6)	Identify alternatives that may contain a solution. Examine items on the alternative list using quantitative tools such as spreadsheets, fishbone, Pareto, etc.	Identify criteria to: weigh, prioritize, validate, and calculate risks & consequences – methods that will filter (alternatives 4) towards a solution + contingencies for implementation & tests (6). Note: Filtering variables come from 3DMM's Executive plans: Objectives (O2), Strategies (O3)	Use various tools to test values for best-measured implementation. These sets may also appear in terms of Budget, Reports, project mgt GANTT or scientific laboratory procedural results	Use various tools to re-examine the level of success of implement plan (#6) to ensure reliability, prevent faulty events, and seek additional change level requirements, re-engineering and re-design, etc.
Example 1: Need a scientifically based theory/model that will help answer	Example 1: Create several scientific models that	Example1: Determine which one of the scientific models answers best all of the	Example 1: Use scientific model to test for qty and qlity value,	Example 1: Use scientific model to address up to 5 levels of tests

and predict data. Example 2: Need to buy a vehicle within my budget, least maintenance expenses, at least 20,000 miles/year. I must purchase transportation within the next two weeks.	contain the data, from which it will be possible to make predictions. Example 2: List various means of transportation that will allow one to arrive on time.	scientific data evidence, predictions. Example 2: Identify which mode of transportation means meets the criteria values – set in #3 'definition' and test in #6; have it quality guarantee (#7)	performance standards. The hypothesis must reflect progress through stages of 'uncertainty.' Example 2: test-drive the right vehicle type/ brand.	/change management; Example 2: Procure quality-optimized transportation. Or, alternative: move to new location closer to productivity hub.

Today's common error is to have the Filter #5 pre-define everyone one of the DMPS 7 step. Such an approach is not scientific but an ideological approach. I call this the 'Luxury Syndrome' because it reflects a situation where an decision-maker (individual, organization, ideology), begins with a wish item – identifying a value e.g., a' Luxury vehicle' at stage #5 – the Filter instead of #1 or at least #3 (Definition). By beginning with stage #5 (Filter) the decision-maker pre-defines all stages of the DMPS - #1, #2, #3, #4, #6 and #7. Here is what happens: at #1 – the luxury-syndrome decision-maker fixes, 'reduces' the knowledgebase to a 'Luxury vehicles.' At #2, psychologically disqualifies all alternate views. At #3 pre-defines the transportation requirements 'gap' to a luxury vehicle. At #4 eliminates all alternatives that vary with the luxury vehicle option and lists quality evidence as reasons, but actually become excuses.' At #5 selects the obvious subjective 'fact' – the 'Luxury item.' At #6 this 'fact' is tested for implementation, and it is discovered that the decision-maker will near bankrupt him/herself on operation and maintenance, but since the luxury decision-maker has pre-designed the 'scientific' tests at his/her 'Filtering'#5 stage, the 'tests' 'facts' yield expected 'prediction' – all other results are considered to be 'contaminated,' 'inconclusive,' and 'undeterminable,' i.e., 'uncertain.' Since so many of these tests remain 'uncertain,' and this 'uncertainty' is not processed through the DMPS stages (#1 through #7), the Filtering #5 process acquires a 'close to reality and truth' description. Such luxury decision-makers and problem-solvers inevitably acquire a false sense of infallibility, like those Scientist/Astrologer/Wizards of Babylon. They have redefined the language and the 'scientific/ideological' process to such an extent that uncertainty associated with astronomy once again blends with astrology; chemistry blends with alchemy/spontaneous combustion; and complexity becomes nothing more than a fast hand movement 'now you see it, now you don't.' Note in Table 4, Filter #5 'Luxury' pre-defines option A and uses other alternatives as de-prioritized 'excuses.'

Table 4 - Tabular List for Alternatives/Hypotheses Formulation (#4)

Alternatives	Qualitative Data per Month				Filter (#5)
Options	Cost $	Maintenance	Gas & Toll	Total	Qualitative
A1 ◀-.	300	200	60	560	.- Luxury
A2	320	210	60	390	Business

A3	280	100	60	440	Wk Mobile
A4	290	50	40	380	Family Car
B1	230	30	30	310	Motorcycle
C1	100	10	0	110	Walk

'Luxury Syndrome' is where Filtering (#5) reduces all alternatives to A1 option. Others options=excuses.

Similarities between the Hypothesis and Theory

What are the similarities and differences between the hypothesis and a theory? The *ECID* book provides up to 12 points of similarity between the hypothesis and a theory. The difference between the hypothesis and theory hinges on the theory's affinity to law-based requirements. The Theory is located at stage #7 – Quality Guarantee. The hypothesis is located in the Alternatives #4 location.

Hypothesis (Gk: hypotithenal – is to put under, to suppose.) is the process of specifying the phenomenon (-a) through quantitative alternatives analysis using the necessary analytical tools whose results will be filtered: weighed, prioritized, examined through criteria for prediction, tests, quality analysis and implementation. Since the hypothesis reflects 'uncertainty,' it is subject to and tracked through the DMPS stages.

Theory (Gk: thea – a view + horan – to see) is a supervisory view or a model that simulates plausible sets of facts, phenomena, processes and rules that reflect or are based on existing or potentially discoverable laws. The theory will provide decision-making and problem-solving means for short- and long-term quantitative analysis (filtering), whose results and applications will be verifiable, and validity checked. Since theory reflects 'uncertainty', it is subject to and tracked through the DMPS stages.

Figures 9, 10 and Table 4 can be compared with Figure 11 for comparison of DMPS with the Uniformitarian scientific method. Later, Figure 25 will compare DMPS, Uniformitarian and Creation filters.

Figure 11 - Comparison: DMPS vs. Uniformitarian 'Scientific Method'

Decision-Making & Problem Solving Method (refer to Figures 9 & 10)	Uniformitarian/Evolutionary Scientific Method
#1 Start with a Knowledgebase (KB) = 3DMM (Figure 5). KB + standards = *'contingent certainty'*. This includes not only quantitative and qualitative empirical data and information, but also laws, standards, rates, measures, principles/procedures of economics and mathematics	The 3DMM is not mentioned. Note: uniformitarian past is 'naïve realism' and an extrapolation of current processes, rates (see #3 below). It provides a 'hypothetical' history and 'extrapolated' knowledgebase. Hypothetical is still within the realm of 'uncertainty' and not within 'conditional certainty.' Uniformitarians reject many parts of Christian Catholic knowledgebase (3DMM).
#2 – *Uncertainty* – Psychological phase (see Figure 8)	Filtering process (step #5) pre-distinguishes and qualifies only the uniformitarian vs. the 'subjective'/'religious' methods. Uniformitarianism seeks a rationalist/empiricist certainty, which is re-enforced by relativism & reductionism.

#3 – *Define* – describe gap in terms of desired solution (#6)	Begin with today's (current) condition, processes, and rates. Extrapolate into past, to identify potential cause to current conditions (using simple to complex process, reductionism, relativism). Definition guided by Filtration (#5)
#4 – *Alternatives (tools, quantitative)*. Formulate a hypothesis and predictions researching missing yet probable data and quantities. Describe the hypothesis, being within 'uncertainty,' in terms of DMPS phases. Use alternatives to formulate predictions (extrapolations, consequences) and interpretations.	Hypothesis formulation to explain phenomenon – make predictions. Hypothesis is usually confused with Theory. Under uniformitarian influence, Alternatives are 'reduced' through the Filtration #5 parameters.
#5 *Filter* (priorities, weighing, criteria, consequences, impact). These are derived from the 3DMM's Design Executive plans (purpose, objectives and strategies).	Filters test for uniformitarian principles or 'naïve realism.' These are *objectives* and *strategies* [but exclude *Purpose* (law, identity, scope)] derived from the Supervisory level: Policies, procedures, rules & regulations. As mentioned above, uniformitarian Filters pre-determine the psychology (#2), definition (#3) and alternatives (#4), and applicable tests (#6)
#6 – *Implementation* (test, standards, codes) lead to conditional certainty (#7)	Test only for uniformitarian realism (UR) results. Consider other test results as being non-compliant - 'contaminated' 'inconclusive' and 'irrelevant.'
#7 *Quality assessment/guarantee*. Verified certainty, may become the foundation of a Theory (lawful scopes), or falsification	Theorize but exclude DMPS phases, which are used to emerge from 'uncertainty' – thus theory remains at the 'uncertainty' level.

2.4 - Dynamic 8-Point DMPS - Scientific Method

The seven-stage *Decision-Making and Problem-Solving* (DMPS) method can better be understood within two additional 3-dimensional DMPS processes Figures 12 and 13.

Similarly, it will also become clear to what degree the debating scientists use the original scientific method. We should remember DMPS is part of the Design Management Supervisory plans in 3DMM.

The 'DMPS Diamond' (Figure 12) has proved its worth in many other fields: a) industrial management development, b) project management, c) in language learning; d) information consolidation in all types of learning. Below, the content of Figure 12 reflects and helps explain the DMPS and the Scientific Method.

Following this explanation, Figure 13 provides an 'in-motion' application of DMPS. Both Figures 12 and 13 lead the process from an initial conditional state of certainty, and with the introduction of uncertainty, the DMPS/Scientific method resolves the path to conditional certainty. Both the DMPS and the Scientific Method are designed to resolve 'uncertainty' and provide a solution for 'conditional certainty.' It also becomes clear that both 'uncertainty' busting methods do not reflect a linear process.

For example, the scientific method's steps begin with 'observation,' yet upon closer examination we see that 'observation' 'A' when placed in dynamic motion, becomes dependent on, or is networked with 'History'(D) and 'Standards' (E). Both are knowledge-base features they compare the new (unknown) with the known phenomenon (history), measurements and standards. An observation - 'data' or phenomenon (A), is not automatically set into a 'framework' or structure (B) and qualified as 'information.' Phenomenon (A) is compared with D and E and only then is it identified as actually being 'new' information (C). It is these foundational steps that determinate the definition (#3), the structure and information (B) in Figures 12 and 13.

While seeking alternatives decision-makers use numerous 'contextualizing' 'formatting' and 'modeling' tools (B), for example the familiar flow-charting, tabular forms, fish diagrams, to sophisticated scientific, computerized and simulation models and tools that allow the decision-maker to collect, adjust, compare, deduce, induce, infer and predict data & information.

Figure 12 - DMPS Diamond and the Scientific Method

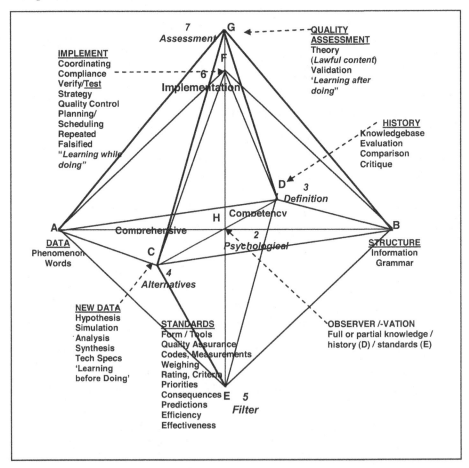

All three (evolutionists, creation and ID scientists) actually begin with the _present_. They recognize existing (today's) processes, evidence, rates. Here they observe, describe, quantify, repeated evidence - life cycles, adaptations, fossil record, and scientific documentation of scientific research in the multiple fields. The differences appear almost immediately. Using Figures 9 – 13 we find that the uniformitarian scientific model begins to filter (#5) and at #1 – provides an edited knowledge-base. At the second stage #2 Uniformitarians qualify only uniformitarian scientists and disqualifies

Figure 13 - The 8-Point DMPS, Scientific Method

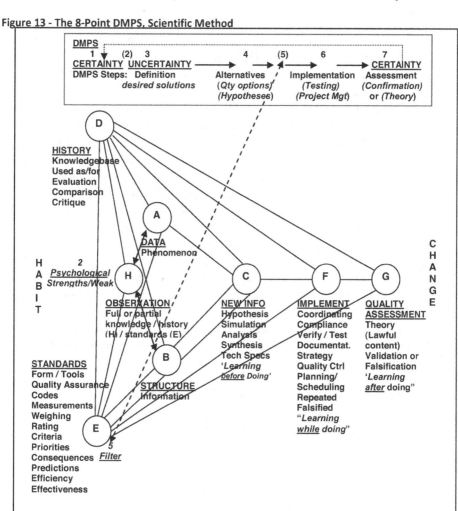

all others – specifically Creation and ID scientists even though the latter two may have better qualified scientists. Then Uniformitarians pre-select and pre-define #3 - all empirical data and information excludes anything that falls outside the 3-point uniformitarian/ evolutionary scope (e.g., dating, geologic column, gradual incremental statistical progression). Then Uniformitarians limit alternatives #4 – again limit alternatives to the 3-point uniformitarian and algebraic scopes. As seen earlier, this type of filtering is established upon the supervisory plans (D3, D4 and D5), which bypass the executive purpose (law, identity and scope) (D1); re-prioritizes objectives (D2) and strategic (D3) plans; and after filtering (#5) whatever is left is then 'tested' for application (within the 3-point uniformitarian and algebraic scopes). Inevitably, much of what is tested is found to be 'contaminated' 'undeterminable' 'inconclusive' or is 'forced' artificially into compliance with uniformitarian predictions or removed from consideration.

In contrast, the Creation scientists use Figures 9 – 13, <u>not</u> from the Filter (#5), but from the historical knowledgebase (#1). These scientists go out of the way to discover and ensure the accuracy of scientific (DMPS) - historical knowledgebase (#1). In #2, Creation scientists use all the scientific output produced by all qualified scientists (including that of the more honest and professional Evolution, as well as, the Creation and I.D. scientists), whose data, information and predictions are verifiable. Creation scientists then define #3 and describe all empirical data and information with the intent of closing the uncertainty gap. They then examine quantified alternatives (#4) - ways of arranging and quantifying this empirical data and information. The filtering process #5 (scientific model) uses the historical knowledgebase (#1) to help filter (#5) the alternatives. This is done to seek what explanation and predictions can best explain the empirical scientific data and information. In addition, which alternative can be best tested (#6) and yield verifiable and reproducible predictions.

The Intelligent Design approach is similar to the Creation scientist's approach with the difference that: 1) the initial knowledge-base (#1) contains laboratory produced data and information; 2) the filtering process (#5) is not conditioned by issues of origins, fossil records, taxonomies, etc. instead, it is designed within micro bio-engineering, sub-chromosome information processes and cosmological constants scopes. One then hypothesizes this complexity within a teleological view.

2.5 - 'Objective and Unfettered Scientific Method' (OUSM)

The reader will recognize many features of the OUSM since these are in part evident in many of the Figures (9 – 13, 16, 18, 22 among others); the '12 limitations of the scientific method' (see chapter 2.6), and leaves the question of origins and ages to the last phase (5th) of the scientific process interpretation.

The great debates among the Evolution, Creation, Intelligent Design, and Hybrid Uniformitarian theorists demonstrate that there is something more to the scientific issues than purely empirical evidence for lab testing. The most obvious incorporations into the investigative process include: a) whether the system is open or closed; b) assumptions for date empirical data and evidence; and c) what constitutes scientific evidence, data and interpretation. It is not so much empirical evidence as the three incorporations that are the subject of the debates' controversy.

The most evident dating method relies on direct methods: a) corroborated evidence of objective, record of sequential events, conditions, structures or artifacts; and b) indirect methods that provide interpretations of fossil record, carbon 14 test, radioisotope method, calculation of outflow sedimentation, tree rings, strength of the Earth's electromagnetic field, speed of light, etc. Here we find assumptions (filters) that reflect philosophical, ideological, and legally historical purposes, objectives and strategies. Yet, one would anticipate that there might be a knowledgebase of scientifically validated /corroborated evidence, a repository of purely objective and unfettered scientific information - a data warehouse of objectively assured data/evidence. This scientific data warehouse would include not only the a) operational layer for sourcing data; b) data access layer with tools to catalog, load; c) metadata layer where a dictionary for warehouse and data accessed, retrieved, for analysis reporting ; d) information access and analyzing tools; but also include 'business'/scientific intelligence tools. One would be able to scrutinize the tools, protect empirical data from premature dating and aging interpretations.

1st OUSM level - Empirical.

The empirically objective and unfettered scientific method (OUSM) can start when for example a master competency qualified and certified scientist comes to a steep wall in the Grand Canyon. This scientist - who may be a geologist or paleontologist - observes the rocks or fossils within sedimentary strata, measures, weighs and identifies samples some of which positioned in multiple layers of various thickness and composition. Most of the strata lie horizontally, while others curved, angled upwards, or even appear positioned vertically across multiple horizontal layers. Among these, the scientist may find fossils, artifacts, fossilized footprints situated in different strata.

The scientist simply describes, records, photographs what he/she observes. The scientist takes samples, makes plaster copies, measures, places reference marks, uses *various techniques* of 'dating' the samples, etc. The scientist carefully records (documents) all of this. Having done this in one location the scientist proceeds strategically to repeat the same process in many other representative areas. The scientist will eventually find the source and the outlet of this Grand Canyon cavity. The scientist notes that throughout the length of the Canyon there is a small river running at the bottom of the Canyon, which he/she observes, measures, weighs samples, photographs, etc. Finally, the scientist writes a scientific article for publication in a science journal. He/she may even venture to create a geologists' tourist brochure; create a video for the National Geographic or contributes much of this information to a school textbook. No interpretation or dating provided at this stage. The 'dated' samples mentioned above are part of cataloging objective sample records. Dating may be an outcome that may be considered in the 4th phase of the OUSM process.

At this procedural and standardized data collection level, will it matter whether the fully qualified/certified scientist who performs empirical analysis is an Evolutionist, Creationist, Intelligent Design, Moslem, Buddhist, or ancient Greek? Not really, as long as the scientist does this empirical work faithfully, and others can verify, confirm and duplicate what has been documented or data warehoused. In other words, scientific empirical objective and unfettered work is valid.

2nd OUSM level work: - Comparative Stage.

Scientist or groups of scientists search and get reports from the knowledgebase – data warehousing, to compare all available empirical evidence in totality. It is an empirical evidential database – a global view. These scientists may 'go high tech' and view evidence produced through satellite world scanning and deep-sea submarine excavations. Scientists coordinate and correlate this geological data and information. They find that most of the Earth's strata either isn't layered horizontally in a systematic A, B, C, D, E and F fashion, but vary - B, A, E, F, D or F, A, D, C or B. Some layers, e.g., A and F, are missing altogether in most areas; or supplemented with very different ones G, H and I layers. In addition, there are single strata that go cross-continent, break off at the seacoast of one continent and continue across the ocean on another continent. Scientists discover that most fossils are marine creatures. Scientists discover six to twelve square mile areas that may consist of massive graves of intermixed fossilized species. There are polystrate-fossilized trees and fossilized skeletons that span across multiple layers of horizontal sedimentary rock deposits. Volcanic lava flows reflect mega-volcanic activity that reaches up to 25+ times greater spans than anything in that is in our recorded history. They note that there are up to 50,000 volcano craters around the Earth that appear to have comparatively limited weathering features. Scientists calculate that it would take 15 million years to weather all continental landmasses into the oceans. They identify numerous human artifacts in massive coal beds ('out of place artifacts'). There are massive layers of crude oil deposits that are still under extremely high pressure. They note that massive and rapid land uplifts formed mountain ranges. None of this evidence is 'dated' at this second OUSM level work.

Scientists examine the surface of the Moon using coordinated high power telescopes, satellite photography, deep scanning tools, and man lands of the Moon. Scientists examine rock samples brought back from the Earth's Moon. Robot space ships bypass or land on other planets (e.g., Mars, Venus, and Saturn) and examine conditions there. Send out Voyager spacecrafts to the edge and beyond the solar system.

This same type of comparative process is done in other fields. Biologists examine the multiple living creatures (plant and animal), create taxonomies to distinguish their specificity and compare these mammals with those identified in the fossil record. Scientists note many similarities between existing and fossilized creatures. One difference is that in many cases fossils tend to lean towards gigantism. There are also differences. Many fossil species do not have counterparts among those living today (e.g., land, aquatic and flying dinosaurs). There are rare finds also - some of the dinosaur marrow and flesh had not fossilized, and mammoths are found frozen completely and intact with buttercups between their teeth. There are fossilized animal and human footprints in most rock layers. Other scientists use the latest super high tech electron and atomic microscopes to discover ready-made sophisticated programmed information systems and nanobot super high technology at sub-chromosome levels. The nanobot operations appear to be on 'automatic' – thus demonstrate engineering software that covers robotic operations, maintenance, repair, manufacturing, self-reproduction and replication. This is a nano-technology that is a 1000 years ahead of our time. This includes all electronic, information and robotic codes and standards that boggle the human mind.

In comparing and cataloging this knowledgebase, it is important that all this investigation allows other scientists to verify, confirm, falsify, as well as, repeat what the first scientists discovered and recorded. This creates an accurate knowledgebase.

3rd OUSM level work: Identify Alternate Causes.

Where OUSM's first stage is the empirical stage of data collection, the second stage is comparative analysis of the Knowledgebase, the third stage helps create and determine alternate causal relationships among the data, information and knowledge in the knowledgebase/data warehouse. Quantitative and qualitative alternate models allow for an analysis of strengths, weaknesses, consistency, standards, and quality of data.

In other words, using Figure 13, scientists re-examine Data (A) and the various way it fits within the Structures (B); test the validity and completeness of the initial Observation (H), to Filter (5) this through Standards (E) and Historical reliability and consistency (D) to identify New Information (C). This may require the updating hypotheses (C) for re-filtration (#5) in order to be able to make better predictions, or make better applications (F – testing, implementation) e.g., discovering crude oil; rates or reforestation; the comets' appearance, frequency, life cycle; role of the thyroid gland in the human body.

Re-examination will no doubt identify a mix of confirmed and missing information. Confirmed new information will be cataloged - updating the knowledgebase. Depending on the set priorities, additional research will be designed and scheduled to fill some gaps, formulate hypotheses (alternatives) where data/info is unavailable. Also the process will allow one to check if such additional research can help clarify theories, which are law-based implementation or guarantee standards. At the DMPS quality guarantee, (#7) phase examine if a theory may explain multiple unknowns. The condition of both the hypothesis and the theory are tentative and therefore reside in an 'uncertainty' status. Being unknowns the hypothesis and theory are subject to the DMPS phased analysis, planning and scheduling. As mentioned earlier, the hypothetical and theoretical uncertainties are tracked through DMPS phases.

Thus far, we have constructed an 'unfettered' scientific – Knowledge-base with contingent certainty. Scientists accurately and transparently have recorded objective information, filtered it twice for accuracy, and laid out opportunities for further advances by refining the accuracy of the knowledgebase, i.e., distinguish between uncertainty and quality assured contingent certainty. The knowledgebase's objectivity is further confirmed by its compliance with the '12 Limitation of Scientific Method.'

4th OUSM level: Application of 3DMM

All scientific activities conducted during OUSM's three stages demonstrate, above all, that these processes are managed. As such the fourth step helps examine how this OUSM functions within the 3DMM – see Figure 14 - the 6th item of the '12 limitations of science.' The 3DMM will include the models of thinking (Figure 15); Five types of qualitative change management (Figure 16); production process (Figure 17); gap depth determination (Figure 18); geometric reality (Figure 20); Qualitative infrastructures and energy densities (Figure 22). It is only after these 3DMM examinations that scientists can introduce methods for identifying dating options of evidence. One can reference dating only within the content of a scientific global knowledgebase that complies within the 12 Limitations of Science (see Figure 14).

The reader can measure and then score the differences among the debaters (Evolution, Creation, and Intelligent Design) by examining the extent and areas where the debaters compromised, replaced, miss-defined or complied with the 3DMM knowledgebase. More specifically, to what extent have the '12 Limitations of Science' been breached? To what extent has original science (i.e., pursuit of accurate knowledge) and the scientific method (DMPS) been maintained?

2.6 – '12 Limitations of Science' – Scored Comparison

Figure 14 provides the means for measuring compliance - 1, or deviation – 0, on the12 limitations of science. The score then provides the percentage for compliance with the 12 items. The higher the score - the more the debater promotes objective science, reduces uncertainty, maximizes objective conditional certainty, and is transparent with his scientific method.

With the subtle redefinition and re-interpretation of science and the scientific method during the past 200 years, very few scientists and philosophers of science now see the

Figure 14 – '12 Limitations of Science' - Scored Comparison

#	12 Limitations of Science	Evolution	Creation	I.D.
1	Does address origins	1	1	0*
2	Value corroborated with historical record	0	1	0*
3	Variables correspond to formulae	0	1	1
4	Proper use of tools, systems & procedures	0	1	1
5	Apply original scientific method	0	1	1
6	Application of 3DMM	0	1	1
7	Four to Five Mental Models	0	1	1
8	Five change levels of management accounted for	0	1	1
9	DMPS – scientific method	0	1	1
10	Ideological implementation of assumptions	1	1	1
11	Distinction between DMPS and uniformitarianism	0	1	1
12	Scientific labs used to reduce uncertainty	0	1	1
	Total – Score card result:	2	12	10
	Corresponding %	17	100	83

* I.D. does not officially address origins and pre-history, although individual members are free to believe what they wish.

limitations as well as the opportunities that these two options suggest. In the past, scientists were well aware of the process fragility in the 'pursuit of accurate knowledge' (science). This pursuit threaded through a forest of uncertainty and either led to greater conditional certainty (knowledgebase), or got many thinkers and scientists to lose their bearings, orientation and balance – thus maintaining 'uncertainty.'

There are many modern and historical definitions of the scientific method. The scientific method that led to our modern time must have contained features that the contemporary modernist view now obscures. However, by using historical tools one can identify the magnitudes, content, distance and dependencies evident in science and the scientific method. This history allows us to identify the original scientific trail. Both science and the scientific method reflect the following stages:

a)   Start with a knowledgebase that contains law-based quantifiable standards and codes that manage conditional certainty (3DMM)
b)   Use investigative techniques, tools, and procedures to identify specific areas of 'uncertainty' – both internally and externally to the knowledgebase (a). These investigative techniques are the quantitative and qualitative decision-making and problem-solving methods (DMPS).
c)   Conduct the decision-making and problem-solving (DMPS) process to help convert specific uncertainty to conditional certainty – i.e., improve and quality guarantee the accuracy of the knowledgebase (a) 3DMM.

The Figures and Tables described in this book help clarify these three stages and contribute to a *Science Score Card* that will help identify who among the debaters is most scientific.

Because definitions of modern science and the scientific method contain many ideological imports, it is necessary to identify at least 12 limitations of science (Figure 14) and the scientific method. Both books detail each of these limitations:

Limitations 1/12 -, scientists cannot empirically examine or test *origins* of the universe. Whether they postulate a Big Bang, Steady Stage or Divine Creation, they will need to re-construct such origins through indirect and secondary sources. In such cases, scientists:

1   Create scientific models (e.g., evolution, creation, ID, hybrid uniformitarian)
2   Use available empirical data and information
3   Make assumptions (values, hypothesis, theories, filters – e.g., uniformitarian, legal historical-based)
4   Make predictions of what #2 is likely to look like, if any of the #3 is valid
5   Origins' definitions (assumptions)(#3) should help explain the framework of empirical data, information or knowledge (#2). Scientific models use empirical data, and when properly extrapolated should meet scientific predictions.

Limitations: 2/12. Short of speculating about origins, scientists may attempt to *reconstitute or reconstruct past* or *historical conditions* and *events.* For example, ice age, catastrophic conditions, consequences of large meteoric impacts, change in species, etc. Scientists can 1) begin with current existing geologic, paleontological, biological, inter-/stellar evidence, as well as, physical laws and/or geometric natural laws; 2) confirm historical data (geology, archeology, artifacts, and pyramids) to extrapolate to, or reconstruct past conditions.

Scientists may extrapolate from existing conditions and may identify singularities – i.e., unique conditions that occurred once in the past and are not likely to occur again. Singularities require one to describe and prove working pre-/during and post singular conditions. However, such unique conditions may not have existed. If they did, original conditions and tools may no longer yield to scientific theorization. One may then attempt to construct a workable framework and then test it scientifically? However, singular event(s) potentially compromise evidence for pre-singular conditions. What are the tools, conditions and evidence that scientists would use to access the non-existent condition? Uniformitarian may postulate a singularity in form of their Big Bang theory, but as it will be shown below, such a condition, which is based on algebraic subjective rules leads to absurdity (see Figure 23). The Creation Scientific Model provides legal historical documentation for the existence of three singular events that had affected the universal scale. In this case, there are at least two issues to resolve: 1) the required nature, source, condition and accuracy of the historical legal documentation. For example external corroborative evidence (archeology, geology, paleontology, demographics), and b) integral features (geometric natural law, geometric solids, infrastructures, stylistic devices, etc). In this Creation view, these must provide conditions and events that existed during and after each singularity leading to predictive conditions that exist today (qualitative energy densities, rates, cycles and infrastructural changes). In this case, since most of the current scientific evidence flows from the last (third) singularity, it will be necessary to account for a) an initial super-mega change or impact that would be, b) followed by secondary mega self-adjustments. On Earth, such initial and consequent self-adjusting changes remain observable, describable, testable and reproducible evidence - geology (hydro-tectonics and global sedimentation), paleontology (fossil graves of mixed species), evidence of mega-volcanic effects, climatology (ice age), condition of ocean floors, demographics, etc.

What Are 'Facts'?

Limitation 2/12 must also address the issue of what constitutes 'facts' since the evolutionary debaters have emphasized the axiom of 'facts' and that the fact of evolution is almost certainty. One feature of facts is that they link to processes and frameworks, which themselves define the fact(s). This is evident in Figure 13 where 'data' (A) links to: D-history, E-standards, before it reflects its B-Structure. An example for this is considering why parallel lines, that are crossed by a perpendicular line, forming 90 degree angles form different 'facts' or 'errors' when placed in different geometric frameworks – Euclidian (linear), Riemannian (spherical) and Loboshevsky's (horse saddle). Similar fact-framework relationships appear when the five levels of qualitative change management emerge (see Figure 16) – similarly appearing facts on different infrastructural levels are significant different from each other.

Here is a list of frameworks and examples of how a fact is determined within a framework:

A   Assumptions may be viewed as a 'leap of faith,' but actually reflects the Executive (D1-D3) row and appears as a Filter #5. Example:

'History progresses from simple to complex'
'History shows a gradual deterioration of conditions'

B   Interpretations provide a strategic (D3) or tactical approach – that test alternative(s).

'The Earth's geologic strata measure millions and billions of years'
'The Earth's geologic strata reflect results of recent global hydro-tectonic catastrophic events, and secondary longer-term mega-adjustments

C   Evidential trends are used to project historical, existing or hypothetical factors

'Today's conditions, rates and events are measures for those in the past
'Conditions, events and rates are unstable, have fluctuated dramatically in the past.

D   Observations described, measured and quantified:

'Paleontological evidence demonstrates change of one species into another under challenging environmental conditions' – macro-evolution.'
Paleontological evidence demonstrates creatures fossilized in the Earth's sedimentary layers show that both have been formed in recent relatively

| short periods. |

E   Descriptive – provides a workable formula or explanation based on a limited number of data:

| 'Speed of light is part of the electromagnetic spectrum that travels at a constant rate through a vacuum and whose path may be bent by gravity. |
| 'Speed of light is part of the electromagnetic spectrum whose speed varies, may be bent by gravity and may degrade with the medium through which it travels. |
| 'Gravitation can be explained by observable behavior of objects that move within its medium.' |
| 'Geometric fields/laws (e.g., Riemannian, Lobachevski) reflect non-degradable structural frameworks' (see Tom Van Flandern, 'The Speed of Gravity – What the Experiments Say.' Meta Research, Published in 'Physic Letter' A250.1-11 (1998); http://Idolphin.org/vanFlandern/gravityspeed.html |

F   Inferred – interplay of cause, rule and effect in a situation where only two of the three are known:

| Deductive: find effect when cause and rule is known.' |
| Abductive: find cause when effect and rule is known.' |
| Inductive: find rule when effect and cause is known.' |
| For millions of year comets have been originating in the Kuiper Belt/Or cloud located beyond the oribit of Neptune at its furthest track. |
| 'Comets are of recent origin that proves that the Solar System is young.' See: Pioneer 10 and 11 spacecrafts did not detect the existence of a Kuiper Belt beyond the Solar System's edge.' See the latest link: http://news.yahoo.com/s/space/20080702/sc_space/voyagespacecraftrevealssolarsystemedge. |
| 'The Grand Canyon was caused by the works of a small river working over millions of years.' |
| The Grand Canyon was caused by catastrophic two-week drainage of three large lakes. |
| 'Random statistical and reductionist rules explain origins and workings at micro-cellular levels.' |
| 'Engineered workings of cellular flagellum suggest a design that cannot proceed through random statistical means and can best be explained in terms of a designing supernatural intelligent designer.' |

G   Analogy – comparative coherence mapping and alignment of source (s) (original), target (simulated) and network (links) conditions and processes that meet same purposes, objectives and strategies.

| 'Results of convergent evolution represented by homologous structures – the living tree that proves common ancestry' |
| 'Taxonomy, a knowledge-base that catalogs, for example, existing and fossil species by various groups and subgroups, such as: domain, kingdom, division/phylum, class, order, family, genus, species and other sub-divisions. Not how one species macro-evolved, speciated into another.' |

These examples demonstrate that the concept of 'fact' lies on slippery ground - clearly, a framework can help convert a fact into an illusion. Scientists must ask the following question: 'what are the processes, contexts and conditions that surround 'facts'?' From the above examples, facts reflect assumptions, interpretations, extrapolation, tactical observation, description, inference, analogy and other. Facts may be materialistic, ideological, religious or mythological. For example, although the spherical view of the Earth had been known from the earliest times, humans have still found it compelling to construct and successfully run empires on the practical 'fact' that the earth is flat. In this case, the source, target framework linked to experience proved to these users that this fact was observable, quantifiable, testable, and verifiable every day – they used Euclidean geometry. During the same historical time, there were also those who used spherical geometry to navigate around the world (Carthegians).

Similarly, throughout history, in a purely naturalistic environment, people have tracked the life cycle - from birth to death. This was an 'observable and testable fact.' They noticed that children in one family differed from each other in a number of ways: physically, temperamentally and in interests ('observable fact'). Husbandry has shown that animals and plants, through selective breeding, brought about preferred traits ('observable fact'). It was also noticed that several animals resembled each other in many ways but were of different species (observation, description, fact). By extrapolating from these 'facts' some determined that with enough variation among the species (cladistic taxonomy) a specie family could change to become another specie (assumption, inference – not observable fact). These observers became theoreticians when they assumed a second layer of extrapolations into the past. They formulated that life emerged through spontaneous generation -e.g., past assumptions of maggots appearing out of rotting meat, crocodiles morphing from dead tree trunks at the bottom of the lake, or under the right chemical interaction - within the 'primordial soup,' life emerged from non-life (assumption, not fact).

Early Christians were aware of these popular 'factual' views. Christians recognized existing contradictions among these pagan 'facts' and what Christians had in their Christian knowledge-base (scriptures) and holy/sacred traditions. In this knowledgebase, there were all types of divinely identified priorities. These defined facts that pertained to all key subjects: laws, statutes, judgments, a document of case histories, precedents, design of types, identification of syngameons and kinds. Similarly, it was noted that the Tabernacle's design had multiple applications: management, project management, psychology, government, social services, international relations, rulership (e.g., King David and Kingdom of God), statecraft, diplomacy, priesthood, authority, patriarchal responsibilities, decision-making problem solving –i.e., solving uncertainty, and the clear identification of and definition of what constitutes certainty. This knowledge/ wisdom-base encapsulates multiple, interlinked information, physical project time-lined events, and physical laws - engineering, agriculture, seasonal, economic, health, etc. Here, it is unique to find the extent to which the Christian Bible - in contrast with all other philosophical and cultural mythologies - has recorded historical events in a specific pattern that resembles the details in the format of the Tabernacle in the Wilderness (described above). We find that Noah's Ark contains key pieces of information that reflect engineering design that have to overcome extreme cataclysmic environmental conditions, temperature inversions, ocean currents that are significantly greater than those existing today. Shipbuilding continued for 120 years on the highest geographical location from where three rivers flowed, yet this ship was nowhere close to any oceanic body of water. This third millennia B.C. engineering design measurements reflected those of oceangoing ships that had to overcome catastrophic climatological and hydro-tectonic forces whose dimensions are inconceivable even today. This ship was to survive more than 12 months of navigation, while carrying the most unusual zoo cargo imagined. The logistics, catastrophic details that can be correlated today with geological, paleontological and demographic scientific evidence and investigation is superbly studied by John Woodmorappe 'Noah's Ark: A Feasibility Study' and others.

To deal with 'facts' it is necessary to identify the DMPS stages as part of the historical and legal 3DMM (knowledgebase, conditional certainty – see Figure 13).

Limitation: 3/12 - Variables within the Formulae

Scientific models can express conditions in terms of *mathematical formulae*. Relational links among mathematical data and variables can make this to be an extremely sensitive tool. Scientists may correctly or through error include redundant or exclude from consideration vital variables or data. Such is the case in the example of Newton's Law of Gravitation. This gravitation equation, constructed solely on the work function – Work = Force x Distance, excludes consideration for energy. To ensure that energy would never come into play, the formula underwent further subtle change over time becoming: $W = Fd \cos(\theta)$. Energy here becomes non-existent. This approach: provides a simple description of work and not of gravitation itself; does not explain what gravitation is; closes opportunities for finding gravity's energy source, as well as, does not identify gravity's rate/speed, which originally appeared to be faster than the speed of light. Einstein's theory does not clarify this issue either because, in part, Einstein capped the speed of light (electromagnetism) at an artificial rate – a constant. Yet, laboratory tests have accelerated light beyond and below this light's constant. In addition, scientists have reduced light's speed to one foot/second, as well as, slowed it down to zero. Furthermore, light exhibits bending qualities (speed variations) while passing from vacuum through other media (e.g., liquids), and through gravitational fields, or disappears in black holes. Similarly, measurements show that some galaxies move away from each other at speeds greater than the speed of light – suggesting that space itself distorts and moves galaxies apart.

Here are some comparative methods for establishing timelines, 'clocks,' and dating methods.

Uniformitarian/Evolution scientists use several methods that are supposed to provide long age time measurements. Among these are:

a)   Geologic Column, with its zoo of fossilized animals and plants, that represent a multi-billion year 'clock'
b)   Rocks dating through the uranium-based measurements, potassium-argon, and argon-argon methods
c)   Radioisotope dating
d)   Short-term organic carbon 14 method
e)   Light travelling through space from distant stars and galaxies, used as a constant to confirm and prove long ages

The Creation scientists, who look at the same data, provide an alternate interpretation to these age facts. The Geologic Column does not actually exist. Geologic strata are not actually set uniformly in an A, B, C, D, E, F strata order, but is more often mixed like decks of cards: B, D, F, A, E, C or other variations with many letters missing or doubled. Similarly, with fossils – 80% of fossils are marine animals. Here and other fossil layers, do not reflect the age of the strata but rather provides evidence for the speed at which the creatures have been able to escape a calamity – the slower creatures were buried first, while the faster and stronger were buried in the following layers. The sedimentary layers and the fossils in them, demonstrate the speed at which the creatures have been able to escape a calamity – the slower creatures buried first while the faster and stronger ones buried in the following layers respectively. Therefore, these sedimentary layers are not a reflection of time, but of flow and sedimentation speed, weight and accumulation. This view is confirmed by evidence of similar strata where 'younger' fossils are located three to six layers below 'older' fossils – areas as large as 28 square miles. We find mass graves – up to six square miles in diameter that contain different species of fossilized animals mixed together that allegedly lived eons apart. Therefore dating through sedimentary layering and fossils are not time related.

Evolution scientists derive their various uranium-based measurements for dating on several layers of uniformitarian assumptions:

1.   Initially, the rock contained only parent isotopes (U^{238}) and none of the daughter isotopes (Pb^{206})
2.   (Pb^{206}) is the result of decaying parent isotopes (U^{238})
3.   The rate of decay had been constant without being leached by water
4.   Additional U^{238} had not entered the rock sample from outside sources. This formula ignores the hydrogen escape rate.

However, when all elements of the formula are properly aligned in the formula, the unfettered science presents a different picture. The hydrogen escape rate alone provides ages in the 1000's of years and not billions of years.

With similar uniformitarian assumptions, it had first been predicted in the 1960's that there would not be any radiogenic argon (^{40}Ar) in rock such as basalt and volcanic rock when it was formed, yet 'non-zero' concentrations of $^{40}Ar^*$ (or excess argon) were found in dozens of places around the Earth.[x] Details of this text continue in *ECID*.

Various explanations are given for such evident disparity and 'isolated' aberrations, but it also becomes clear that the radioisotope dating methods, due to the apparent discordance of dating output are far from yielding reliable absolute ages. This is evidence for formula minimalization. Uniformitarian scientists did not consider other more plausible tests. They did not recognize verified explanations offered by establishment scientists who tested conditions that 'accelerated' radioisotope decay. Alternate views strayed from the uniformitarian guidelines/rules (see USA's 'National Academy of Science' below), and the establishment shelved these results.

Scientists increasingly present better alternate theories that test assumptions. For example, data and processes that describe mount St. Helen's volcanic activities. These demonstrate that alleged 'long-age' sedimentation, fossilization, landscape deformations can occur in hours not hundreds of millions of years.

The significant shortcomings of the Big Bang theory are evident.[xi] Theorists have re-explained the details of this theory countless times to the extent where the human language has become a barrier – e.g., things (something) of all sorts appear out of 'nothingness.' There are separate types of time – pre-/during/ after the first mega-explosion of the cosmic egg? There are processes that contradict physical law. What is the progress of mega-explosion stages during which time, space, energy, gravitation, movement were forming, and when the electromagnetic waves, proto- and actual light (electromagnetic fields) emerge and behave at each of the five qualitative stages? Billions of years (our current years) pass in nano seconds. These eventually slow down to seconds (of our perceived time) in order to condense and then expand again – leading to our current condition of the expanding universe? All this occurred within our notions of the speed of light while everything floated through 'dust' that the sun had not yet absorbed. This dust should have resulted in a theoretical accumulation of between 1000 ft to 233,000 ft on the various solar systems' bodies. Yet such cosmic dust remains undetectable in any form anywhere on the Earth, Moon, or any other Planets. See more about the Big Bang below and its direct derivation from the philosophical and algebraic model – Figures 19 and23.

At the same time, there are more than a 100 dating methods that suggest drastically reduced time scales. Here we may list: sun's shrinking rate, sun's emission of gamma rays; zircon crystal's age - determined to be less than 10,000 years. Such crystals contain known lead leakage rates. What about Earth's slowing rotation; the diminishing electromagnetic field around the earth? The RATE Project and many other dating issues that raise serious questions about mainstream science's assumptions. Clearly, scientists must revisit all of these formulas and assumption, considered within the objective and unfettered scientific scope.

Limitation 4/12 - Proper Tools, Systems and Procedures

The capacity of scientific tools, systems and procedures that function within a framework set within *limits and tolerances of accuracy*. This is perhaps the single reason why scientific progress had been relatively slow in the past. Observation and measurements were limited to the instruments and mathematics used. For example, how would Darwin's theory of evolution change had Darwin had access to super high tech electronic microscopes? Clearly, Darwin's 19th century organic model of thinking did not allow him to conceive the overwhelming complexities of the nano-robotic information processes at the sub-chromosome levels. From the 1980s, scientists discovered that every living cell (plant, animal, human) contains infinitely precise data and information processing (built-in software, hardware codes, processes and standards) that is further magnified by nano-robotic engineering operations, routines and programs. Closer examination reveals that these processes provide evidence for self-inspection, quality performance, self-maintenance, self-repair and self-replication. How can statistical progression (evolutionary chance) in a challenging environment account for this multi-infrastructural management system?

Similarly, in the dating process, if scientists should exclude a single variable from the rock dating process, this would result in dramatically different dates. Such is the case with procedures involving radioisotopic dating. Formulae that *include* the helium escape rates receive results in the 1000s of years instead of formulae that *exclude* the helium escape rates and thus receive results in the billions of years. Alternatively, when the Apollo mission returned with Moon rocks; test results helped: falsify the existing four theories on the formation of the Moon and placed a question mark on the value of the current dating methods. For example, lunar geologists discovered examples of large lunar areas with 'younger' dated rocks located inside 'older' dated rocks.

The condition, methods and locations that scientists measure, depend on their strategic approach. This is their 'assumptions' and the tools, systems, and procedures that they use.

The uniformitarian mind-set dismisses a global flood scenario. They replace their perceived 'myth' with another uniformitarian myth. For example, uniformitarians hypothesize up to five major ice ages that lie upon three types of evidence. Evidence based on: geology (scouring and scratching of rocks, glacial moraines, drumlins, valley cuttings and the deposition of tillites and glacial erratics)[xii]; chemistry (oxygen isotope ratios in sediments, acid concentrations, sedimentary rocks, ocean and ice sediments)[xiii]; and paleontology (where and how fossils are distributed away from cold areas). This evidence is to substantiate up to five major ice ages coupled together with over 30[xiv] intermediate and local ice ages. However, the same evidence substantiates a single recent 500-year ice age that occurred between 2000 and 1500 B.C.

Table 5 - Hypothesized Uniformitarian Ice Ages

#	ICE AGE	PERIOD	YEARS AGO	SEVERITY
1	Huronian	Early Proterozoic	2.3 billion	
2	Snowball Earth	Cryogenian	1 billion	Most severe
3	Andean-Saharan	Ordovician & Silurian	400-430 million	Minor
4	Karoo	Carboniferous/Permian	250-360 million	Polar ice
5	Glacial, Interglacial	Pliocene	2, 58 mil to 10K	N. Hernis

Scientists proposed various uniformitarian causes for these ice ages. Evolutionists admit that these are highly controversial views. They attribute several causes: quantity of CO_2 in the Earth's atmosphere; volcanic activity; position and movement of continents, which affect ocean currents; changes of the Earth's orbit[xv] (Milankovitch cycles)[xvi]; the Muller and MacDonald's three-dimensional orbit inclinations;[xvii] and c) the Sun's energy output.

Creation scientists identify processes that few uniformitarian scientists suspect. Evolutionists have not yet described, tested causes or identified Ice Ages mechanism that must account for at least seven key catastrophic pre-requisites. These would leave specific signatures upon geological, chemical and paleontological evidence that uniformitarian theories do not address. For an Ice Age to occur there must:

1 Be massive volcanic and plate tectonic upheavals, which
2 Leave massive amounts of volcanic aerosol in the atmosphere that contribute to a vast temperature drop by reflecting solar radiation back into space, but continue to
3 Heat the oceans – this
4 Increases evaporation - a 10 degree Centigrade in air-sea temperature difference, with 50% relative humidity, will evaporate 7 times more water at a sea surface temperature of 30 degrees Centigrade at 0 degrees Centigrade[xviii].
5 As volcanic ash covers the Earth this contributes to the cooling of the atmosphere
6 These interactions (1 through 5) cause increased precipitation (X 50 magnitude of that existing today). These conditions have been simulated on the CRAY computer and,
7 These conditions must persist for at least two consecutive years.

With these minimal conditions, in less than 500 years after the worldwide Flood, in an environment where up to 10,000 active volcanoes heated the earth, oceans and the air, increased ocean evaporation that converted to continuous rains and black rain at equatorial zones, while snow would accumulate rapidly in the Polar Regions and on higher mountaintops. In the lower valleys, particularly within the vicinity of the warm and hot oceans, temperature would remain temperate and land would support abundant plan and animal life. Further regions would succumb to sudden glacial weather patterns. This would explain conditions where we discover hundreds of thousands of frozen mammoths in the northern regions of the Earth. No doubt, similar freezing conditions existed on the Antarctic continent.

After the volcanic activity declined on the Earth:

1. Cloud cover dissipated
2. Oceans cooled enough
3. Volcanic dust cleared out of the atmosphere, causing
4. The atmosphere's temperature to rise, causing
5. Within 7 to 200 years a significant melting of ice packs around the Earth
6. Raising ocean levels between 500 ft at the equator to 100 ft in higher latitudes

Scientists successfully simulated these conditions on a CRAY computer (Los Alamos Laboratories) based on geologic, chemical and paleontological evidence. The software helped reconstruct conditions and helped predict causes and consequences within the context of a single post Global Flood Ice Age on the Earth.[xix] Recent drilling in the Greenland and Antarctic Ice Sheets revealed that Ice Age durations can be measured within 1000's of years instead of 100,000s to billions of years[xx] as suggested by uniformitarians.

The proper use of tools, systems and procedures – is key in determining the accuracy of knowledge (science). One factor that held back science throughout history had been the lack of proper tools, systems and procedures. Since the renaissance tools, systems and procedures have been opening new doors. Yet today, after having witnessed the mega-software and nanobotic evidence at sub-chromosome levels, it becomes evident that we have only begun to road to developing proper tools, systems and procedures. Contrary to what uniformitarians (evolutionists) propose (see NAS's rules on 'complexity'), the proper use of tools, systems and procedures must not be restricted if we are to help uncover new empirical data. These tools and data cannot be hidden, diverted or distorted. There is much drilling in the arctic and Greenland ice and ocean bed floor, the Creation scientists have purchased many CRAY computer hours to simulate data and forecasts the various critical weather conditions.[xxi]

Limitation 5/12 - Application of the Original Scientific Method

What is scientific? During the past 200 years, all kinds of things appear within the parameters of what is to be scientific. Not only do we have the simple to the overly complex definitions proposed by philosophers of science, textbooks, but also various ideological concepts founded upon economics, dialectical materialism (DIAMAT, HISTMAT). Here we find that Evolution is science according to what courts have decided, it is 'almost certainty and truth' while laboratory procedures find an inordinate amount of 'contaminated,' 'irrelevant,' or 'inconclusive' results. Some classroom subjects (e.g., biology, geology, and physics) reflect simplistic scientific methodological linear steps guided by the uniformitarian filter. These exclude any science that does not conform with this uniformitarian filter (see Figure 13). Some evolutionists give cursory credit to pre-Darwinian works of the ancient Greek philosopher-physicists who sought natural causes and processes, and a few who have contributed to the rationalistic-empiricist cause – e.g., the mechanists Descartes and Newton – the result however clearly aims at suggesting that most of pre-Darwinian science is fundamentally less than scientific or un-scientific.

During the past two hundred years Uniformitarians have mounted a concerted effort at identifying a 'demarcation line' between the 'sciences' and religion.'[xxii] Eventually, among these philosophers of science the theorist Paul Feyerabend discovered that history points to an eclectic approach to the scientific method. Feyerabend discovered that scientific reach has reflected five trends. Scientific research had not been based exclusively upon logical or methodological rules of science that were distinct from sound reasoning. There was no special scientific authority to draw upon. Those scientists at various times had violated every scientific procedure. That science is inseparable from the larger body of human thought and inquiry; and in practice and principle, science had not entirely been empirical (e.g., see the various 'pure' scientific theorizing that goes on – e.g., string theory). Although many scientists, theorists and historians questioned Feyerabend's discovery[xxiii] Feyerabend's research documentation still stands unsurpassed. Upon closer examination, this scientific investigative process becomes true if the proper science and the scientific method reflect management principles (3DMM and DMPS) and environment. Here, Feyerabend's discoveries of alleged inconsistencies in science and the scientific method become consistent within the original historical process of scientific development. It is today's ideology focuses, i.e., uniformitarian filtered scientific process ('Luxury Syndrome') that appears too rigid and reductive.

The Young Earth Creation scientists succeed much better in the area in the application of the original scientific method. They accept a historical rather than a synthetic view of the 3DMM; recognize a definition of science that includes all stages; and respect the full range of alternatives filtered through executive plans applied during the decision-making and problem-solving (DMPS process).

Dr. Henry M. Morris, the father of the modern Young Earth Creation Scientific Model, has successfully rescued the original Christian Biblical scientific model by providing it with strategic re-positioned Creation Scientific Model. This allowed the Creation scientists to fight a new battle with the stealthy uniformitarian marketing promoters.

The twentieth century's two World Wars have reshuffled the 'scientific' deck of cards in world politics. Although the West had successfully neutralized most effects of two extreme versions of the application of the Darwinian Theory (Marxism and Nazism), the West itself had succumbed to the new atheistic myth, which aimed at destroying the last vestiges of Christendom from which the modernist scientific method emerged. It is the uniformitarian theses that helped achieve these gruesome results. Dressed in scientific garb, the uniformitarian initiative set itself to remove all true science from the civilization's thinking. Where science originally meant 'pursuit of accurate knowledge,' the uniformitarian's science removed, or has re-written the history of the reliable 3DMM. This uniformitarian scope is to provide a synthetic past by starting and importing existing conditions, processes and rates into the past; extrapolate materialistic causalities by adhering to the simple-to-complex, primitive-to-modern processes.

Thus, uniformitarianism re-interprets, relativizes and redefines the process of the scientific method:

1   Standards (see 'E' in Figure 13 - the 8-point DMPS)
2   History (D), data (A)
3   Structure (B); observation, perception, awareness (H)
4   Scope of quantitative alternatives – C
5   Methods of implementation, testing (F), and
6   Quality assessment/guarantee (G)

Uniformitarians (evolutionists) will continue, as any professional cultist would, to implement the next step:

7   Provide the uniformitarian tool by which the individual, family, society, institution, organization will daily brainwash its/themselves of all past 3DMM 'conditional certainty' through reductionism and modernism. Thus, by cutting all connections to original historical and standard roots, modern man will become putty and servant within the New World Economics.

One may ask, how successful have the uniformitarians been in achieved their goal? The answer is that by the year 2000 AD, all mainstream Christian organizations had drunk from the glass of uniformitarian wine. In the process, they have become pantheists by having converted to Theistic Evolutionist, Progressive Creationists, and have become ardent contributing members in the Modernist Ecumenical Council (MEC). With such pantheism, original Christendom, from which emerged the quest for contingent and absolute certainty (accuracy of knowledge, the scientific method and geometric natural law) has now 'reduced' itself to becoming 100% proof uniformitarian. Within such an environment and thinking process, the Book of Genesis and other parts of the Christian Bible have simply become a literature with symbolic, metaphoric and mythological content rather than a documented history set in legal format. Should the ancient Biblical writers have used the uniformitarian modernist method to document this legal history, they would have 'adjusted,' 'reduced' or 'brainwashed' themselves to the nose-to-the-ground naïve realism that we see today – the production of a myth.

During this migration to uniformitarian realities, the now 'Modernist Christian Church' (MCC) has forsaken its history, traditions, the qualitative knowledgebase 3DMM and the true tools for getting out of uncertainty to certainty. Instead, MCC relativizes its language, and straps itself for a ride that includes the 'Luxury Syndrome,' cynical reductionism – i.e., cultism and a subjective 'naïve realism.' Dressed in this new robe the MCC qualifies to play a role on the pluralistic marketing stage. Where during the historic past, Christianity had been a faith that could not be made to fit into Caesar's Pantheon of cults; today, with the generous helping of the Modernist heresy, MCC not only fits comfortably in Caesar's Pantheon, but also plays a leading role among the Pantheonians.

One would ask, what was the original Christian scientific method designed to achieve? Original Christianity has had a history founded upon geometric natural law, genealogies, executive-based covenants, commandments, statutes and judgments, precedents; social, family and health laws and regulations; state and court law and protocols. This 3DMM knowledgebase contains the formula for establishing the Kingdom of God, and contains the earmarks for identifying the Kingdom of Babylon.

Geometric natural law when properly understood contains global engineering management plans; purposes, objectives, strategies, with engineering and infrastructures designs; life-cycles comparisons; rules for husbandry, selective breeding, medicine, property and economics management, maps, records of Kings' management, precedents and records. This record also contains a detailed history of singular events, global cataclysms, and world droughts, documents civilizations populated by giant humans, and civilizations and conditions that existed before and after singularities whose effect identified in today's geological and paleontological formations.

On the other hand, the uniformitarian naïve realists and cynics may 'reduce' this reality to 'stone age' record keepers, to the realm of mythology, but by this very act, the modernist uniformitarians disqualify themselves from discussing issues that pertain to geometric natural law (e.g., Riemannian physics and mathematics) and issues of sub-chromosome information and nano-technology. Specifically, they disqualify themselves from considering themselves as authority on issues pertaining to the executive plans and stylistic levels of the 3DMM since they reject these axiomatically.

On the other hand, by rejecting uniformitarianism, the 'Young Earth Creation' scientists and their science publications remain within the framework of historical definition of science. They reflect all applications and uses of scientific tools. They successfully, verifiably and reproducibly move uncertainty towards conditional certainty[xxiv]. When *uniformitarian scientists* attempt to analyze creation-catastrophic interpretations of empirical data, this is done on purely *uniformitarian* grounds and definition of terms. Uniformitarians do this within a narrowed scope (reductionist) of alternatives, peppered with 'Luxury Syndrome' logic, and by wearing sun shades that make them either overlooked much empirical data or categorize this data as being 'contaminated' 'undeterminable' or 'inconclusive'. What uniformitarian critics present are nothing but uniformitarian 'naïve realism' positions. Clearly, the uniformitarian approach is not justified upon the original meaning of science. The high-level debates clearly demonstrate this. For example, examine Mark Isaak's 'Problem with a Global Flood,' Second Edition, The Talk Origins Archive, November 16, 1998 – http://www.talkorigins.org/faqs/faq-noahs-ark.html ; and the rebuttal by J. Sarfati, 'Problems with a Global Flood?' 1998, http://www. trueorigin.org/arkdefend.asp.

During the past 30 years, all debating groups had been repositioning themselves, adjusting their strategies and resetting priorities. This is particularly true with the advent of third players into the debate – the Intelligent Design group; and more recently the Neo-Theistic Evolution groups: Progressive Evolution and Old Earth Creationism.

Limitation 6/12 - Application of 3DMM (Figure 5)

This sixth interpretative method reflects the best that the 3D Management Model details. Certainty in the 3DMM is inversely proportional to the 3DMM's 27 plans, infrastructure's content as it reflect the 'Objective and Unfettered Scientific Method' and as it reflects geometric natural law. A quick review of the first five limitations examined above, demonstrate a systematic approach to overcoming scientific gaps. Here, scientists in its empirical stage, without assigning dates and ages, gather, document and test empirical data, compare data and information accuracy that they collected globally and now reside in the knowledgebase. Scientist re-examine the knowledgebase to identify gaps, areas of 'uncertainty,' provide quantified alternatives, earmark areas for potential hypotheses (i.e., alternatives #4). They prioritize and test all alternative's data, information and knowledge for application, which lead to 'conditional certainty'. In the current sixth stage, scientists test for quality and 'ascertained/ absolute certainty'.

3DMM managers interpret data, information and knowledge set within mental models, geometric natural law, change management levels, etc. within OUSM. This approach views all evidence within a systematic management scope, to ensure that knowledge is accurate, contingently certain or quality guaranteed. Data, information, knowledge links among management plans, infrastructural levels, and vertical columnar directional, processive and info-base. This data, information, knowledge is verified, falsified, validated, confirmed internally, environmentally, as well as within the hypothetical and theoretical realms – which reflect 'uncertainty' and are subject to the DMPS.

As within industrial project management, it is necessary to answer questions of ethics, attitude, culture (*management style* S1-9). These answers are favorable to/conducive with scientific research. Under such circumstances, it will then be easy to contrast answers that emanate from outside the unfettered scientific system. Unfettered scientific results, within a 3DMM, would allow for triangulated answers that seemed until now to have been unanswerable until now. Within approach, the 3DMM would warehouse all infrastructures, 27 plans and other networks that would link information and knowledge, to answer the best supported issues such as -

• 'What is the origin for this entire marvel or existence?'

• 'How did all this come about and where is it going?'

• 'What is the proper behavior and attitude for enhancing knowledgebase certainty?'

• 'What is the nature of a favorable environment that ensures long-term accuracy process?'

Such questions would not be restricted to reflections of the uniformitarian 'Luxury Syndromes.' Valid scientific questions continuously 'pursue accurate knowledge'. This is why these questions are located within the 3DMM among Executive plans (purpose, objectives, strategy) on the Management Design level, and specifically on the Management Style levels. This would clearly show the sub-value of the uniformitarian approach, which by definition eliminates all executive level questions and specifically the Management Style infrastructure. Due to the uniformitarian agenda discussions of such issues within the scientific context appear to be unusual today.

Events that began in the 16th century challenged Western Europe Christendom in an unsuspecting way[xxv]. A radically alternate view emerged that questioned the validity of the foundations of <u>all</u> Knowledgebase systems and of their contingent certainty. This new emerging approach during the last 500 years had:

a)   Truncated the original knowledgebase system. It eliminated and reassigned executive plans at every level in the 3DMM to the supervisory level. This approach 'reduced' awareness of the management style plans and through this, promoted principles of 'naïve realism.'
b)   Placed all foundations upon relative conditions rather than upon the DMPS that led from uncertainty to conditional certainty. This inevitably established a continuous state of 'relative uncertainty' rather than attempting to achieve 'conditional certainty.'
c)   Actively pursued an antagonistic strategic replacement policy in all aspects of its perceived rival's approach (Christian) to its 3DMM.
d)   Re-coined the vocabulary, concepts, standards and source of authority in the host culture – stressing subjective, relative, substitution and diminishing value
e)   Reversed objective and unfettered science by violating the scope of the '12 Limitations of Science' thus causing an implosion of the 3DMM. Through this, it accelerated the dynamics of failed civilizations.

Limitation 7/12 –The 3DMM and the scientific method (DMPS) reflect four to five basic *mental models* (Figures 15 and 20-22). These five models help manage the hypothesis, theory and have predictive value. For example, the Newtonian and the Cartesian models of thinking reflect *mechanical* and *reductionist* forecasting methods. The Bergsonian and Darwinian models reflect an *organic* model of thinking. Karl Marx's

'Das Kapital' reveals the economics *processive* model. The Intelligent Design model reflects the *information* technology and engineering model of thinking. The Christian Biblical or the Creation model reflects all of these mental models allocated to their corresponding content and researched for all of their proper application.

People can hazardously misuse mental models, particularly when they focus upon un-contextualized 'facts' and 'evidence.' They easily may hamper or arrest scientific development. Scientists can arrest science unconsciously or through premeditated policies. Such an approach leads to absurdity - see examples on 'facts' (above).

Limitation 8/12 – Five Change Levels (Figure 16). The engineering industry recognizes up to five qualitative change levels that describe more than a simple linear change. These change level concepts contain qualitative infrastructures. Sometimes a seemingly simple change process, at a different infrastructural level, can be a qualitatively different change.

Figure 16 details five change levels are: 1) static or copies, 2) change or adaptation, 3) improvement or re-engineering, which also includes supervisory systems; 4) innovation or re-design - purposive executive systems, and 5) invention or third level change (e.g., hypothesis of the hypothesis).

The uniformitarian / evolutionary concept, in order for it to achieve scientific value and conditional certainty, in areas where an alleged *mi*cro-change becomes *ma*cro-change, the process must describe and test potentially up to five levels of change. Both Evolution and Creation may agree on change that occurs at 1), 2) (adaptation) and possibly 3) within an inherited pre-programmed code. However, evolutionists leave the scientific domain when they attempt to explain higher *ma*cro-change levels – 3), 4) and 5). Here, evolutionists must address change at several infrastructural and supervisory systemic levels – an area where statistical and linear methods cause havoc. Without having this requirement met, evolutionists revert to generalized morphing concepts such as alchemy and spontaneous generation. To avoid this situation, evolution scientists must describe processes, management tools and the challenging environmental conditions that would lead to qualitative (*ma*cro-) changes at the infrastructural, supervisory, functional and executive levels as well as columnar directive, processive and info-base areas. All of this must include information software, data warehousing and protocols, as well as, nanobot technology management that contribute to countless decisions making – i.e., internal DMPS and 3DMM.

Limitation 9/12 - DMPS (Figures 9 thru 13 see related texts). The comprehensive *decision-making and problem-solving process* (DMPS) process is the foundational tool in the scientific method. This staged process helps convert uncertainty to conditional certainty. Here also, scientists categorize the hypothesis (alternatives) and theory (quality guarantee) stages. History records various types and stages of the scientific method development. Recently, neural systems, data-warehousing and business intelligence systems have facilitated this scientific investigative process that lead to finding specific solutions that meet objectives and purposes. Others allow wider creative options, exploring quantified alternatives and hypotheses.

Limitation 10/12. Scientists use '*assumptions*' when they construct scientific models. This helps interpret data and make predictions. Fair-minded scientists recognize that assumptions are 'biases' - value judgments forced into the scientific investigative process. Scientists and philosophers view this bias as a disguised hypothesis or theory. Assumptions are essentially DMPS procedural filters. In management, such assumptions reflect executive objectives, strategies and must include purpose. Scientists may begin from any assumption, but should not allow this assumption to violate the original intent of the scientific process – such as to eliminate or distort empirical data by narrowing the scope (see item #12 below)

This section examines a blurring line between ideology and science. Groups promote ideology as science, and then *im*properly define both science and the scientific method. They mis-define, substitute, miss or overly simplify the steps of the scientific method. Proponents of such a 'science' usually accuse others of not 'understanding' their 'true' science. In this process, they continually modify the process as they 'explain' the 'true' meaning of this imponderable (relative, reductionist) 'science'. For example, school student may ask 'what is a trilobite?' The packaged answer will include: a) an oversimplified description – 'a prehistoric sea creature.' The teacher then couples this with, b) an ambiguous 'uncertain' date - 'that lived million/billion years ago'[xxvi]. The astronomical date implies a reality that existed beyond the 6000-year Biblical Creation option. Such answers affect the student's dignity, integrity and self-determination (see Figure 5; 3DMM's Stylistic level plan S8). This formulation sets up an invisible barrier to any further exploration of alternatives (see #4 in DMPS). In other words, such questions and answers remain in a state of 'uncertainty' 'relative' level – a path that closes investigation towards 'knowledge accuracy' and 'contingent certainty' (stages #6 and #7 in the DMPS process). Therefore, students simply memorize scripted, prescribed 'scientific-sounded' formulations and reproduce the same in order to pass the 'science' test. In the outside world or on the job, unless the student, employee, manager, clergy or any audience speaks such uniformitarian magic language, the individual, group, society and civilization, it can be concluded, that such entities:

1.   Cannot be scientific or objective
2.   Remain un-programmed, reactive and uneducated
3.   Must have dormant or latten cultist leanings
4.   May even potentially harbor Creationist (pseudo-scientific, insane) tendencies
Should anyone doubt this trend, we can examine areas where this has happened - uniformitarian States such as Soviet and Nazi empires. Here, the uniformitarians after having defined their uniformitarian realities began to filter and qualify its populations – non-conformists to the uniformitarian matrix were diagnosed in terms of being psychiatric or counter-revolutionary patients. Similarly, in the United States, few have

noted the trend. Here, professional scientists, university professors and management personnel that do not adhere to the uniformitarian party line, are legally pronounced to be nothing more than 'cultist quacks', and lose their professional positions and reputations.[xxvii]

However, similar conditions exist among those who exercise oversimplified religious awareness. These can get off the true scientific path and can also direct many off the path. Most of today's Creation Science groups are of the Protestant denominational persuasion. Here, it is sometimes difficult to distinguish the fine line between science and faith. With 44,000-plus Protestant Denominations, one would have to search with a candle in daylight the fine line between the original Christian and Christian cultism. All have 'done away with' geometric natural law, holy/sacred traditions, original authority, and parts of the originally canonized scriptures. Here we see that both the uniformitarians and the denominations 'erase' much of Christian history, sing from the same modified historical music sheet – here they identify the same 'historical heresies;' quote the same 'injustices' and 'genocides;' and reduce the cultural language to simple selection of defined trigger words and concepts that form barriers. The majority has strayed from the historical path, their reasoning opens a gap that helps them challenge what is 'faith' and 'science' – since both 'faith' and 'science' have been mis-defined.

Here are some ideological positions from which assumptions are drawn. Some reflect a statement that seems to based on scientific evidence, but such statements begin either with a conclusion or in the middle of an argument. These may imply some interpretation of origins, and draw inferences or preclude a programmed unique conclusion. In other words, we have the 'Luxury Syndrome' exemplified.

Uniformitarian/evolutionists' notion of: Abiogenisis – life emerging from non-life. No scientist has ever proved Abiogenisis in the lab, nor does anyone know how such an event can occur. Numerous hypotheses have been proposed, attempted in the scientific lab without success and abandoned as soon as counter proposals emerges. In the past 100 years, there have been at least 22 such theories of processes that can be divided into five categories:

1.	Historical - spontaneous generation, Darwinian proposed prebiotic soup, and Haldane and Oparin's chemical evolution.
2.	Organic molecules theory (Millers' experiement – the neo-evolutionist's re-engineered Primordial Soup); Deep Sea Vents, Fox's experiment, Eigen, Wachtershauser, Homochirality, Radioactive beach, Self-organization and replication.
3.	Protocells theory (Genes first; Metabolism first, Buggles performance
4.	Other model variations include the Clay theory, Autocatalysis, Deep-hot biosphere; Primitive extraterrestrial life; Lipid World; PAH world; Polyphosphates and Multiple genesis[xxviii].
5.	The most recent attempt had been offered by Kenyon's 'Chemical Predestination' a required text during the 1970's, showed that DNA arranged itself, as if magnetically -ions, into the right protein formations.

In contrast to the evolutionists' Abiogenisis, Creation scientists identified the law of Biogenesis – life comes only from life. The DNA arranges and reproduces all elements of the organism's processes according to pre-programmed 'software'. Life, created by the Eternal Lord God allows life to exist and adapt within its environment, life procreates within the existing, physical, environmental and biological laws. This created life reflects the original type scope and has the internal programming that does not drift beyond its kind. All adaptive change demonstrates a loss of information, while mutations demonstrate an accumulation of 'debris.' Life passes the genetic code and its re-calibration through inheritance. As the microbiologist Louis Pasteur (1822-1895) proved there is no 'spontaneous generation.'

Intelligent Design: the above-mentioned Kenyon, the co-writer of 'Chemical Predestination,' has gone a step further and soon identified the basic flaws in his book's basic scientific assumptions. Kenyon proved that there was not a single chance for having biological life to emerge through chemical evolution or chemical predestination. He showed that such notions are solely established on an information theory. There is more to life than bio-chemical processes. Such theories exclude greater complexities – information technology at sub-chromosomal levels. Along with information engineering at such levels of magnitude, there are also issues of nano-robotic engineering, where robotic systems have capabilities to self- copy, replicate, translate, shepherd other components, self-maintain, self-manufacture, and self-replicate. None of this could possibly emerge through statistical progression –chance events.

Limitations 11/12 – Distinction: DMPS and Uniformitarianism

The *scientific method is not synonymous with uniformitarianism, materialism, naturalism or reductionism.* These false synonyms are assumptions - filters in the DMPS (Figures 9 thru 13).

Dictionaries and encyclopedias provide only an approximate definition of uniformitarianism. Here, uniformitarianism appears as an assumption, concept, basic principle, generalization, or axiom. Most dictionaries associate the term with geology, while other encyclopedias provide an etymological study of the term. In the late 18th century, James Hutton offered the term 'gradualism' to counter the notion of 'catastrophist.' Both terms were to explain geologic formations. John Playfair popularized gradualism, while Charles Lyell (1797 – 1875) incorporated gradualism as 'uniformitarianism in his 'Principles of Geology' in 1830. Another concept expressing uniformitarianism is 'actualism' because the term suggests that it is through the current conditions, processes and rates of change that we can extrapolate any kind of condition that may have occurred in the past – 'the present is the key to the past.' Theorists used this 'deep time' geology, gradualism, and the need for 'observation' (current evidence, processes) to contrast and replace the notion of 'recent time' (Catastrophism - Biblical

Flood). Uniformitarians, recently brought back a limited view of 'catastrophism' in an effort to explain meteoric effects upon a planet. The limited 'catastrophist' ideas was prompted by predictions and observations documented by the Shoemaker-Levy about the nine comets that plunged into planet Jupiter. It suddenly became apparent that conditions of the surface of the various Moons in the solar system, and on many parts of the Earth itself, provide evidence for numerous meteoric catastrophic impacts. However, such momentary and local catastrophes do not extend to supersede the uniformitarian/ evolutionary process.

The uniformitarian view and process is the foundation for the Theory of Evolution and the ideology of materialism. The Theory of Evolution would totally unravel without its Uniformitarian foundation. When someone is so totally dedicated to the Theory of Evolutionist and will stake his/her reputation and career upon it, it becomes evident, whether such person realizes it or not that such an evolutionists is essentially a uniformitarianist and materialist. In this tri-notion, the term 'scientific' is arbitrarily attached to the ideological notion of 'materialism' and not to the 3DMM and its DMPS, which are negated. The uniformitarian axiom generates inferences - concepts of natural selection, speciation, macroevolution, are but extensions and window dressing for the uniformitarian axiom. Therefore, to determine whether uniformitarianism is scientific, one has to compare it with the scientific method (DMPS) - see Table 6 below.

The comparative evidence shows that the uniformitarian axiomatic process violates every step of the original definition of the scientific method (DMPS). Evolution's original conditional certainty is associated with the philosophical position of materialism – an axiomatic position. The 'uncertainty' steps are: a) materialism, b) reductionism, c) and its reactionary view against and the rejection of all non-materialistically established knowledgebases (3DMM).

Evolutionist Realism (ER) somersaults tree times in the DMPS process. Here's an example: bring scientific evidence for 'beneficial' mutations in challenging environments, through this proving qualitative change from micro- to macro-evolution (i.e., emergence of new genes and new information), the uniformitarian /evolution biologists should jump at the opportunity to use the latest CRAY computers to simulate, produce, test mutational simulation and natural selection. If biology evolutionists have already taken advantage of this CRAY computer to simulate and test mutation processes, then this information has not trickled out of the laboratories. However, the Young Earth Creation scientists did create such software and did simulate the effects of mutations for this and many of their own population genetics studies. In two articles – 'Mendel's Accountant: A New Population Genetics Simulation Tool for Studying Mutation and Natural Selection' by Dr. Baumgardner; and 'Using Numerical Simulation to Test the Validity of Neo-Darwinian Theory' by Dr. John Sanford - both authors had studied mutations and natural selection in human genetics, plant and animal breeding and management of endangered species. Both scientists developed this state-of-the-art software jointly with ICR and the 'Feed My Sheep Foundation.' As they describe:

'…a forward-time population genetics model that tracks millions of individual mutations with their unique effects on fitness and unique location within the genome through large numbers of generations….the user chooses values for a large number of parameters such as those specifying the mutation effect distribution, reproduction rate, population size, and variations in environmental conditions.'[xxix]

While uniformitarians/evolutionists make predictions and theorize that cumulative change proceeds from simple to complex[xxx], the 'Mendel's Accountant' software tracks such mutations and helps prove that natural selection does not detect nor eliminate genetic deterioration due to mutations. Instead, these accumulate over multiple generations thus degrading fitness.[xxxi]

The scientific comparative analyses conducted by uniformitarian, creation and Intelligent Design scientists show that the creation and I.D. scientists use the original scientific method, while uniformitarians violate the process of the original scientific method.

Table 6 - Comparison: DMPS, Uniformitarian, Creation Scientific Methods

DMPS (Figs. 9 & 10)	Uniformitarian	Creation
#1 Start with a Knowledgebase (KB) = 3DMM (Figure 5). KB + standards = 'contingent certainty'. This includes not only quantitative and qualitative empirical data and information, but also laws, standards, rates, measures, principles/procedures of economics and mathematics	3DMM not mentioned. Note: uniformitarian past is 'naïve realism' and an extrapolation of current processes, rates (see #3 below). It provides a 'hypothetical' history and 'extrapolated' knowledgebase. Hypothetical is still within the realm of 'uncertainty' and not within 'conditional certainty.' Uniformitarians reject many parts of Christian Catholic knowledgebase (3DMM).	3DMM legal historical knowledge-base; 3 singularities occurring during recent historical periods–a) Divine Creation of reality, time, geometric natural law, infrastructures, life; 2) drop of energy-density one cycle down the universal cone; 3) initial global hydro-tectonic global catastrophe followed by secondary global geologic, oceanic, climatologically and other adjustments. Instead of a quantitative progression, there is instead a steady qualitative regression of structures, comparative phenomena
#2 – Uncertainty – Psychological phase (see Figure 8)	Filtering process (step #5) pre-distinguishes and qualifies only the uniformitarian vs. the 'subjective'/'religious' methods. Uniformitarianism seeks a rationalist/empiricist certainty, which is re-enforced by relativism & reductionism.	Preparation for objective and lawful methodology – framed within 3DMM, conditional certainty and resolve uncertainty through DMPS
#3 – Define – describe gap in terms of desired solution (#6)	Begin with today's (current) condition, processes, and rates. Extrapolate into past, to identify potential cause to current conditions (using simple to complex process, reductionism, relativism). Definition guided by Filtration (#5)	Describe gap in terms of desired solution (#6)

#4 – Alternatives (tools, quantitative). Formulate a hypothesis and predictions researching missing yet probable data and quantities. Describe the hypothesis, being within 'uncertainty,' in terms of DMPS phases. Use alternatives to formulate predictions (extrapolations, consequences) and interpretations.	Hypothesis formulation to explain phenomenon – make predictions. Hypothesis is usually confused with Theory. Under uniformitarian influence, Alternatives are 'reduced' through the Filtration #5 parameters.	Formulate a hypothesis for missing yet probable data and quantities that is to be researched. The Hypothesis, being within 'uncertainty,' must therefore, be described in terms of DMPS phases. Alternatives are also used to help formulate predictions (extrapolations, consequences) and interpretations.
#5 Filter (priorities, weighing, criteria, consequences, impact). These are derived from the 3DMM's Design Executive plans (purpose, objectives and strategies).	Filters test for uniformitarian principles or 'naïve realism.' These are objectives and strategies (but exclude Purpose) derived from the Supervisory level: Policies, procedures, rules & regulations. As mentioned above, uniformitarian Filters pre-determine the psychology (#2), definition (#3) and alternatives (#4), and applicable tests (#6)	Filter (priorities, weighing, criteria, consequences, impact), these are derived from the 3DMM's Management Design Executive plans (purpose, objectives, strategies).
#6 – Implementation (test, standards, codes) lead to conditional certainty (#7)	Test only for uniformitarian realism (UR) results. Consider other test results as being non-compliant - 'contaminated' 'inconclusive' and 'irrelevant.'	Implementation (test, quality, codes) that lead to conditional certainty.
#7 – Quality assessment /guarantee. Verified certainty, may become the foundation of a Theory (lawful scopes), or falsifiability.	Theorize but exclude DMPS phases that are used to emerge from 'uncertainty' – thus theory remains at the 'uncertainty' level.	Quality Assessment. Guarantee. Verified certainty, or may become the foundation of theory (lawful scopes) or falsifiability.

A minor example where management style – e.g., Ethics (S1), Culture (S3), Attitude (S5) and other management plans, play a paramount legal/scientific importance can be seen in the multi-faceted concept of communication. For example anyone who has an in depth knowledge of general esthetic literature will recognize that human behavior emits information continuously about the self. Such communication may be verbal, non-verbal (behavioral), mannered, intended/ unintended, overt, covert, intentional, or accidental. Even no behavior is information. Information comes through movement; may be passive or active; generated by expression in the eye. Information may reflect ethics, cultural, style. Information is also generated by possessions: clothing, housing, cars and tools we select and use. Here, information can be verified through triangulations, repetition. People become walking advertising boards. Some people are aware of this and have refined their 'behavior,' they think proactively and 'act out' the right role and manner for the occasion, or to help achieve specific objectives or have some control over the environment. All of this is language and communication. The Christian Bible states that on the Day of Judgment[xxxii] each of us will be held accountable for all of our communication and information Every day there are thousands of ways that one can do and say things to bring about positive and desirable result, as well as negative ones. No doubt, behavioral communication reflects our awareness, perception and scale of values. All of this translates in the manner scientists conduct, evaluate and improve conditional certainty. Such behavioral scientifically related details are clear in the 3DMM.

This ethics derived communication is something that all Christians must learn. The highest authority commissions Christians to create an environment where they daily train themselves to speak before Kings and Rulers, and learn to manage as the best of them. King David left a historical legacy of his successes and failures, his use of the DMPS and 3DMM. This is part of and contribution to conditional certainty. The same 3DMM Executive and Stylistics plans affect scientists who conduct scientific work. Their ethics, attitude and culture affect the manner in which, and the type of science that scientists conduct. In contrast, when scientists base themselves on the relativistic and 'Luxury Syndromes' - this predictable tendency leads in the opposed direction. As documented, the uniformitarian position distorts what constitutes 'science,' reduce styled scientific communication to irrational, accusatory and irrelevant objectives, as well as include a relentless anti-Creation/Christian. Such an ethic also places Creation and Intelligent Design scientists into a legal category that is equivalent to 'cultists' 'pseudo-scientists' or implies 'insanity' simply for not complying with uniformitarian tenets and authorities.

The Intelligent Design scientists do not use the Christian Bible to reference historical, legal, or definition of a geometric natural law or Covenant sustaining Eternal Lord God. So there is no true 3DMM Knowledgebase with an Ethics-base objective law to provide a foundational management view. At the same time, the Intelligent Design (ID) scientist may have his/her own private (subjective) views about the content and structure of these extra management executive and stylistic plans, but these are not included into the laboratory environment-based Intelligent Design scientific process view and method. The ID view is Teleological at best. However, ID scientists adhere to profession ethics and do not harbor a destructive intent toward other qualified and objective scientists or objective science. ID scientists recognize in some form the basic tenets of 'objective unfettered scientific method.' The only difference is that they identify a hypothetical super-intelligence, as a means to: 1) explain the astounding complexity discovered in the lab; and 2) distance itself from the: a) outdated uniformitarian ideological positions and b) its unethical agenda against Christendom; c) against the historical Knowledgebase (3DMM) and the d) objective scientific method itself (DMPS). In addition, e) to show disfavor towards the uniformitarian involvement into 'scientific social engineering' that led to the sacrifice of hundreds of millions of human lives during the first half of the twentieth century (Marxism, Nazism, Fascism and other underlying ideologies). I.D. scientists f) disfavor actions taken by *American secularism*, where institutions through new procedures and objectives simply reflect the secularism that had been used during the genocidal period of the first half of the genocidal century.

Many observe and document that secular institutions also pass judgment that allows for continued genocides numbering in the multi-million - abortions; and who deliberate on the criminality, legality and sanity of whole segments of the population (Creationists – the majority of the American population). Uniformitarians border dangerous pseudo-science/cultism that is taught in school. These decisions are not made based on the original American constitutional law, but upon the secular uniformitarian and reductionist foundations/secularism – a constitutionally illegal application.

Limitations 12/12 - The purpose of *scientific laboratories* is to provide tools that reduce uncertainty and helps promote and ensure 'conditional certainty'. However, increasingly, scientific lab procedures accomplish the reverse. Here we can commonly find that when predictions are not born out, the lab scientist finds 'evidence' for samples being 'contaminated,' 'irrelevant,' or 'inconclusive.'

It is necessary to distinguish between a) laboratory procedures whose evidence helps reduce uncertainty; and b) procedures that maintain conditions of uncertainty when predictions are not met due to faulty 'assumptions.' In the latter case, scientists may continue to test not to ensure a more precise certainty but to force the tools and evidence to meet specific predictions. If predictions still do not meet expected objectives, then such scientists will suggest that the results are 'inconclusive' due to 'contamination' 'anomalies' 'insufficient data.' This is pseudo-science.

The recognition of these *twelve limitations of the scientific method* forms an integral part of the investigative scientific method. Today, notions derived from exclusively axiomatic foundations, lead some to accept or generate ideas that may - start with a conclusion; begin in mid thought; offer partially developed views, or appear as pre-formulated 'canned' views – whose buzzwords resemble scientific fact. Such an approach reflects uncritically accepted definitions, a bias offered by an establishment's scientific community, processed through 'peer review,' or assured by 'the overwhelming scientific consensus.' Here, although such bodies under normal conditions may provide legitimate objectivity and productive enterprises, in an algebraic environment, however, these professional bodies begin to exhibit symptoms that in management is recognized as 'group think.'

The scientific lab comes in different shapes, forms, means and sizes. The scientific lab must meet different purposes, objectives and reflects different strategies. With labs and models, scientists may examine and test phenomena in areas of physics, chemistry, biology, geology, paleontology and other fields. The scientist measures phenomena for definition, design, alternate use, flows, change, and are made to behave through catalysts, in ways that yield data.

Some labs/models are used to reverse-engineer processes to see if some hypothetical conditions can help resolve issues. For example, Miller and Urey used reverse engineering to create artificial life. They used reductionist techniques – oversimplified apparatus for abiotic synthesis of organic compounds. The system directed varied mixtures of gases through electric charges, to produce many types of organic compounds, immediately trapped and separated. The product turned out to be both 'right and left-handed' with nothing else that would further separate and distinguish one from the other. They made a variety of assumptions and inferences. The ultimate objective was to see how to created basic organic compounds in the lab, thus attempting to show a step towards proving *a-biogenesis* – emergence of life from non-life. Geological evidence however did not support Miller and Urey's assumptions about early Earth conditions[xxxiii].

Another example of territorial laboratory conditions may be setup to determine Tasmanian devil's behavior and change in isolation, such as that conducted at the University of Tasmania, Australia since 1996.[xxxiv] Here, the creatures developed a transmittable cancerous facial tumor disease (DFTD). Because of this disease and its early death, the marsupials 'adjusted' their life cycle (5-6 year-life-span) and their breeding cycle (between 2 and 3 years), to a breeding cycle that began as early as 1 year to allow the females to take care of their young before an early death (2-3 years). However, it was interesting to see how scientists interpreted this phenomenon. University scientists perceived this change as evidence for macro-evolutionary change[xxxv], rather than for micro-evolutionary adaptation.[xxxvi] Clearly, the graduate students' uniformitarian axioms predisposed them to 'predict' only a linear mechanism for all types of biological change – micro- or macro-evolution. Looking at this objectively, it became clear that this uniformitarian position did not allow one to distinguish between adaptive (micro-change; programmed within the genes) and processes designed to overcome qualitative, infrastructural, re-engineering and re-designing macro-change (macro-change). Micro-changes provide evidence for evident loss and not gain of information at each stage. This micro-change approach allowed other scientists to predict the extinction of the Tasmanian devil within 25 years[xxxvii] instead of the appearance of a macro-changed new species. To achieve macro-change it is necessary to demonstrate 'learned' – i.e., new genetic information that would demonstrate required changes in infrastructural, supervisory, as well as provide evidence for re-engineering and re-design into macro-changes.

A difference exists between qualified or ideological scientists. Both graduated from accredited institutes of higher education - the first is a dedicated scientist who pursues his/her effort wherever research leads him/her. The other, is the ideologically motivated scientist who will bend and 'adjust' his/her research so that it conforms to ideological predictions. Both scientists publish their researched findings, but the second one loves his/her ideology much more than his/her scientific research. This second group involves itself in questionable politics and does not blink an eye when it comes to violating the objective scientific method 'OUSM' and DMPS. Their mind attuned (programmed) to the Luxury Syndrome, they recognize the value of uncertainty and hardly ever reach conditional certainty.

Ideological scientists, during their debate, exhibit less than scientifically satisfactory lab results. The list below identifies methods that ideological scientists use. We also find website debunkers who emulate this ideological trade:

1. Incomplete proofs or results (e.g., 2 of 5 factors) accounted for during the presentation or limited test parameters – e.g., talking about conditions on the Moon they identify 'cold' (instead of the 4000F) and 'dry' (instead of existing humidity at poles).
2. Dissymmetrical comparisons - mis-interpretation of data and conditions under which variables function or behave. Here, fraudulent information appears on graphs.
3. Ascribe motives or methods that are not part of the process. Introduce off the cuff questionable cause. Data or events appear via unknown means.
4. Misidentify source of data and processes
5. Compare apples with pears. Perform analysis on what is finally interpreted as being 'contaminated' or 'inconclusive'
6. Doctor up or revise measurements, values and calibration criteria
7. Mud-slinging – nasty pre-/middle conclusions, disrespect, irrelevant information
8. Intentional presentation of misinformation, knowing that the audience would not check
9. Make emotional mountains out of ant-hills
10. Present alternatives that have no bearing on the issue. Narrow the scope of alternatives. Change parameters
11. Quibble about inconsequential issues
12. Allusions to improperly constructed experiments – sources, movement, changes, temperature range behavior
13. Misinterpret content and context. Focus on insignificant points in order to obscure a sound structure or body of evidence
14. Scientific models established on invalid or limited assumptions
15. Use of equations with inconsistent dates
16. Consider only one cause for real or imaginary fault lines, cracks, uplifts, scratches, demarcation line and points.
17. Application of marketing/sales tactics – 'bait and switch' to attract customers by advertising one desirable product only to sell a product that has lower quality at higher price. xxxviii
18. Reject the need to read clear proofs and answers in order to submit faulty conclusions.xxxix

These 18 are but a few examples of false science magnified during many debates. In some policy-making quarters such an approach is mandatory and promoted by the USA's National Academy of Science - see 'Rules' below.

These high-level examples become sustenance for those who lurk in the dark corners of many websites, where 'debunking' hopefuls congregate. These practices are applied in the classroom among the impressionable, as it is evident in the scientific and book review chat rooms. Articulate chat room motivators gets to practice the 'Luxury Syndrome' art; or use syllogistic logic (if A=B and B=C then A=C) as their preferred tool which allows them to ignore myriad of pertinent, causal, inferred or implied context information. Another favorite ploy is to focus and 'debunk', 'reduce' the meaning of each separate phrase, sentence of though, into isolated and dislocated groups separated from contextualizing paragraphs and sections. This approach eliminates the framework value or meaning. In such an environment, science becomes part of the urban legends and soap opera. Contrary to professional logic and examination, the use of an ideologically based language, and winning through intimidation (violation of stylistic management – 3DMM) has become a rule.

In such an environment, we may meet straight A-science students who may explain the general theory of evolution through well-rehearsed scripts. Here we hear about 'early microbes that began to adapt to isolating challenging environments. Such conditions lead to having the same groups of species that are no longer inter-fertile. Having this process repeating itself countless times over billions of years eventually brought about the emergence of man.' During the Q&A session, it becomes evident that the science student uses syllogisms, provides canned answers on every occasion, but it is also discovered that the A-student has never been taught to think in terms of alternatives, or from a third person perspective, nor has he/she been updated by his/her mentors about contingent information. For example, that:

a) This 170-year old uniformitarian 'theory' had never been designed to attain contingent certainty. Alternatively, it was never to be used to test or prove anything through empirical scientific lab methods

b) CRAY computer-based software can simulate mutational quantitative processes and generates statistics that would help disprove many uniformitarian assumptions, but that such scientific information would only be available through Creationist scientific lit.

c) Life forms adapt (*micro*-change) to challenging environments, but do not re-engineer, re-design, nor re-invent themselves (*macro*-change)

d) Genetic change implies loss and not gain of genetic information

e) The process of natural selection is not designed to detect and eliminate accumulating genetic change 'debris';

f) This genetic debris accumulates and contributes to the steady degradation of the species until its extinction.[xl]

Young Earth Creationist scientists, with their Creation Scientific Model seriously challenge all uniformitarian assumptions and evidence. It is particularly worth reviewing the use of language and concepts where uniformitarians, because of their Luxury Syndrome, make numerous logical somersaults and present a truncated view of the historical knowledge base (3DMM) and blind spots.

The 2000-year developing Christian scientific culture had helped develop nuances and 3D legal-based communication. The uniformitarian language, which is similar to Marxist ideological language has 'speciated' into a separate foreign language, that resembles and reflects a body of reversed values, ethics, attitude, methods, priorities and culture – a *2*DMM. The 3-point uniformitarian scientific method has not only significantly narrowed the scope of creative thought, but perpetually seeks some means for polarizing and eliminating 3DMM-based users. One can take as an example from the Marxism-Leninism worldview – whose believers similarly use a dialectical materialist language. It is worth to notice that the uniformitarian democratic notions stand in opposition to the Christian Biblical 10 Commandments and geometric natural law.

The modern uniformitarian practitioner and thinker, similarly as the Marxists, Nazis and Secularists, use the uniformitarian materialistic view. Here: a) the present is key to understanding the past and the future; b) simple to complex statistical (economic) progression; and c) the ultimate purpose is to eliminate Christian foundations and establish materialist, relative and reductionist reality. The perceived uniformitarian principles are to work in the social arena (social engineering), as they are to work in biological evolution. When a student graduates from some secular educational institution he/she must prove and demonstrate an uniformitarian/evolutionist worldview in order to continuously qualify for a job, science lab, education, political, commercial field, and become a catalyst for 'change'. Such employee and catalyst should never propose or make evident an alternate view (e.g., Intelligent Design or Creation science).

CHAPTER 3 – OBJECTIVITY, LIMITATIONS, MODELS & RULES

3.1 – Thinking Models

The scientific method inevitably reflects four to five mental models. The fifth model of thinking is addressed below - Figures 20, 21, 22, 23 and 26. These models expand, specify or limit the DMPS process. Mental models reflect patterns, values and competencies for processing data, information and knowledge and implemented behavior and conclusions. Companies, organizations and individuals may excel with the use of any of these mental models. Mental models, however, become an issue when they are misapplied – e.g., biological mental models used to interpret information-based processes, where sophisticated networks with supervisory, self-re-engineering, self-design and other features exist.

Elsewhere, studies have suggested many different mental models. For example, we have the Jungian archetypes and Meyer/ Briggs personality types. Similarly, philosophies, ideologies and religions reflect how the human minds work, perceive, analyze and organize ideas. Inevitably, we recognize here mental models also.

The mechanistic mind, for example, views reality in terms of machine functions. Here we have interplay of concepts such as parts, components, assemblies, units, efficiency, productivity, inventory, power, gears, preventive maintenance, etc. Rene Descartes and Newton's works are good examples of the mechanical mental model thinking.

The organic model, used by Bergson, Darwin, Hegelians, and Teilhard de Chardin, visualize the world in terms of organic growth, development, life cycles, evolution, environmental and genetic concepts, *élan vitale*, and concepts of simple to the complex.

The processive model presents a systemic approach, offered by the modern industrialists. They may not necessarily know where the process begins or ends, but they focus only on what is logistically necessary. Here, inputs are processed into necessary outputs. The system has quality assurance and controls, feedback mechanisms that allow for 'self' adjustments. They measure this process. They seek to improve and may re-engineer this process for better performance.

Figure 15 - Models of Thinking

MECHANICAL	ORGANIC	PROCESSIVE	INFORMATION
Sum of parts	Interrelatedness	Systemic flow	Communication
Maintenance	Nurturing	Quality control	Copy, adapt, improve…
Quantitative	Qualitative	Quality assurance	Cycle, goal seeking
Manufactures	Organic birth	In-/output, process, fback	Sensory input, feedback
Wear & tear	Life-cycles	Preventive maintenance	Supervisory systems
Rational	Evolutionary	Project Plan	Neural, knowledge
Matter & energy	Bio-chemistry	Resource, time, money	Dignity
Balance	Stability	Optimum performance	Integrity
Algebraic	Relative	Economic	Self-determination

The fourth group represents information processing-based mental model. The scientist or information-robotics engineer recognize a reality that is intricately filled with networks, software, feedback systems, controls, a knowledge base, learning system. This scientist and engineer recognize supervisory systems, robotics, and goal seeking systems. They seek to refine the operations of 'consciousness' that function within a fixed or changing environment.

It is easy to recognize these mental models in every occupation: literature, management styles, philosophy, and ideology/religion. It is also curious to recognize these mental models during contract negotiation, business round table discussions, debates and political agenda. After one (he/she, group, organization, and civilization) selects its mental model, whose model's logic almost dictates the manner for reaching conclusions or solutions? The rules become virtually predetermined. Gaps in communication emerge solely based on the mental model's pre-determined differences. For example, how would each proponent of the different mental models view an organization?

1   Mechanical:
     An organization is an integral structure of operational and maintainable units, components, assemblies and systems. These elements function efficiently and effectively to generate work and products and function within physic—chemical laws.
2   Organic:

An organization is an organism that reflects functional patterns so that it would meet environmental and internal challenges. The organism may undergo life-cycle adaptations and changes while performing specific function(s) in a competitive environment, while functioning within bio-ecological laws.

3   Processive:
An organization is a systemic process with five fundamental functions that convert inputs into qualitative outputs, ensuring standards, controlled feedback, while attempting to limit delays, noise and waste (entropy); while functioning within customer economic laws.

4   Info-base
An organization reflects timed communication, controls and goal seeking system. It implies capabilities for learning, decision-making and problem solving, recognition, anticipation (planning and scheduling). It includes three feedback levels: at the functional (integrating), supervisory (network automating) and executive (consciousness; functioning within neural networks and servo-stylistic laws) levels that allows optimized behavior within an environments

With these views, it is easy to see why some scientific data, information, framework and 'facts' would fit differently and would affect scientific research, processes differently in the attempt to resolve uncertainty. It is clear that using one mental model instead of another brings about different results. For example:

- An inordinate number of 'unexplained' anomalies
- Uncovered 'out of place' artifacts or fossils
- Force the lab technician to prematurely ignore or shelve 'contaminated' samples. Warehouses of such contaminated or inconclusive samples exist.
- Discover and interpret fossils within the geologic strata simply because geologist and archeologists were 'authorized' to find and discover them – e.g. 'Nebraska man.'
- Identify widely divergent or even contradictory results on dating. For example, different dating methods on the same samples from the same location yields dramatically different ages.

These are but a few examples among thousands, when it is clear that the wrong mental model and consequent procedures lead to predetermined conclusions.

3.2 - Five Qualitative Change Levels

All phenomena, data or information fits into some framework. This framework may reflect up to five levels of qualitative change. Infrastructures separate each qualitative level (copy, adaptation, re-engineering, re-design, invention – see Figure 16). Operations at each level have their own unique specificity - standards, codes, software, etc. Each phenomena or the framework itself may function as a catalyst for change within or between infrastructures if so programmed. Figure 16 provides an extremely simplified version of what actually occurs at each level. A seemingly similar simple change within each of the infrastructures will reflect different causalities, links, processes, standards and codes within their respective infrastructures. It is, therefore, a mistake to attribute changes occurring at the 'adaptive' #2 level, to be similar at each of the higher levels since these higher ones must account for re-engineering, re-design, and have self-maintaining, reproducing supervisory and consciousness software and processes. The 3DMM must be viable, self-correcting and be proactive.

The five types of qualitative change management (Figure 16) contain several pre-requisites: infrastructural levels each containing change processes that vary from simple copies to conscious self-change information and robotic systems. As an analogy, few people realize that the simple **copy** machine is in essence a robotic system. Each of the five changes at the different infrastructural levels must account for converting inputs into outputs, containing feedback and programmed activities and functions within strict performance standards, codes - all occurring within controlled internal environments and without external environments. Supervisory and executive systems ensure and guarantee the quality materials input – processing - output and product guarantee. In living systems, this complex process must include self-maintenance, repair and self-replication. Today's technology does not approach anywhere close to this intricacy. When using proper mental models, the Creation and Intelligent Design scientific models can recognize some this operational parameter. However, the Evolution model inserts an axiomatic linear statistical progressive mechanism, within each level and among the infrastructures. Such materialistic and reductionist tools fail to meet the mega statistical challenges that an organic model must tackle.

Briefly, investigators who use Figure 16 will find that the Evolution model, which restricts itself to linear change only – i.e., mutation + families of species + challenging and isolating environment – will never be able to account for the multiple billions of challenging and isolating environments that must exist within the billions of years-old Earth. These billions of environments must exist to account for the billions of families of species that exist today. Such a condition requires the activity of zillions of coordinated mutations at four qualitative change levels, which must also overcome the zillions of error-coordinated mutations at the four qualitative change levels. If all this should occur, as the Evolution Model suggests, it would require that today, investigators would find a significantly less stable cellular and much more fluid physiological structures to accommodate such qualitative change rates. If such qualitative change rates should occur, they would cause havoc at the sub-chromosomal information and nano-robotic levels, since such mutations and qualitative changes would have to occur at speeds faster than that of light if they were to occur within the constrained billions of years. Today's conditions appear to be very dormant in comparison with what has to occur. The Evolution Model does not only propose spontaneous generation (matter to life), but sets itself on alchemy.

Medicine today provides areas of good lab-based progress, but also atrocious mistakes introduced through reductionist UR/ER thinking. See eugenics - designed to 'remove' genetic defects through sterilization, abortion, euthanasia. Performs needless surgery to remove 'vestigial organs;' expounds nonsense about 'junk' DNA; designs useless procedures, curricula; promotes medicine that has no clinical or predictive value, etc.[xli]

Figure 16 - Five Types of Qualitative Change Management

L	CHANGE TYPE	FLOWCHART	OBSERVATION
1	*NO CHANGE* Copy	Environment 1 → 1 ← - - - - - - - - 2nd a copy of 1st	The duplicate is identical to the original design, content, processes, and info. Duplicate functions in a similar environment
2	*CHANGE* Adaptation Re-alignment Micro-change	Changing environment(s) to adapted processes 1 → ◇ → 2 ← - - - - - - Change *Feedback system*	Entity changes within its 'programmed' scope. Entity adjusts in response to internal/ external feedback. Id new patterns to reflect programmed cycled changes, or permanent change.
3	*IMPROVEMENT* Re-engineering Macro-change	1 → ◇ → 2 → ◇ → 3 Change Improvement *Supervisory System* Changing environment requires self-re-engineering – a supervisory system with strategic capabilities	Supervisory function at 2nd level feedback process crosses an infrastructure. It not only deals with original and copies (2) but also re-engineers new standards, codes resources, for improvement.
4	*INNOVATION* Re-design Mega-change *Evolutionary theorists suggest mega-changes to explain data, but do not offer self-redesign evidence.*	1 → ◇ → 2 → ◇ → 3 → ◇ → 4 Change Improve. Innov. *Supervisory* Consciousness Changing in-/external environmental conditions require self-re-design (software + processes + supervisory systems, codes, standards) through awareness of the environment (in-/external), tools, and competitive forces.	Self-awareness (consciousness) 3DMM – modifies re-designs supervisory functions to help overcome real and potential internal /external challenges. This includes, re-mapping, several re-engineering and self-redesign changes within a cultural setting. Standards, codes energy densities.
5	*INVENTION* Breakthroughs Conscious self-change *Evolutionary research never addresses this 'change' even though nature evidences product results.*	INNOVATION 4 → ◇ → 5 Copy Change Improve Innovation Invention Supervisory Conscious Executive	Theoretical re-examina-tion of laws, conscious (design) concepts. Executive faculties: purpose, objectives and strategies. See qualitative cone's energy/density levels – Figure 22

3.3 - PRODUCTIVITY – Functional Plans (#D7-#D9)
Business and Project Management Applications and Other Applications

Productivity (D9) is the output that meets Objectives, and customer satisfaction. It is also a test for efficiency, effectiveness, reporting (feedback system), elimination of 'noise' and delays. Productivity is the product of what the Purpose (D1) has conceived and designed. All intermediate plans in the management design - between Purpose (D1) and Productivity (D9) are the conversion mechanisms that result in the product.

Productivity is, at least, a 4-stage process that must work perfectly under a supervisory system with proper feedbacks (#4) to ensure the desired output (#3). This process begins with quality assured Inputs (#1) – sourcing, on time resource delivery through the logistics system (warehousing, receiving of purchased materials, quality assurance/control, safety, security, documentation, materials management, labor, distribution). Quality assurance (QA) parameters, include from product sourcing (#1) through raw material conversion (#2) to output (#3). Therefore, the raw resources (#1) are converted (#2) into the final product (#3) to meet the specific product objectives - quality assurance (#1), control (#4) and customer satisfaction (#4) (customer needs, wants and delight). This includes post-service evaluations – seeking quality guarantee.

Figure 17 - Production Process

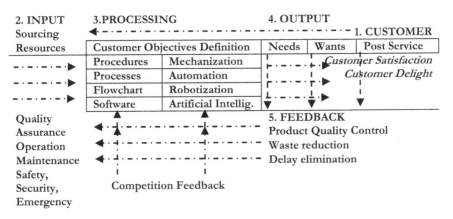

The productivity process includes specific procedures, processes, flowcharted steps, checklists, mechanization, automation, robotization and artificial intelligence. Every step can be automated or genetically programmed within plants and animals.

The productivity process model is a schematic of what also occurs at the information and nano-robotic engineering sub-chromosome level. All scientists must address these issues and particularly the evolution scientists who have theoretically built in a mechanism(s) of mutation and natural selection as the tools of choice to account for the changes that occur at not only one level, but also to all qualitative levels as described in Figure 17. Both the Creation and Intelligent Design scientists do not have to tackle this *statistical progressive concept*, since this concept emerges from the ideological concept – that derives from an organic model of thinking. On the other hand, the Creation and Intelligent Design scientists recognize valid 'copies', 'adaptive' and the properly defined 'improvement levels', but do not have to subscribe to a mere morphing 'statistical progression' to any of the higher three qualitative levels of change. Here, clearly one must consider an external 'pre-programming' approach.

Figure 17, provides another option of viewing how conversion mechanisms work at their most fundamental level.

As mentioned earlier with regards to Figure 16 (Five qualitative change levels), Figures 17 – 'Production Process' and 18 – 'Gap Depth Analysis,' become useful in helping us understand the conversion process that occurs from one qualitative level to another. Infrastructures provide not only a support but also a barrier. It is where quantity is converted to value, quality and vice-versa - where supervisory and executive infrastructural changes occur. These interplay within the Knowledgebase (the 3DMM) – where conditional certainty is established, maintained, updated, standardized and conditioned.

3.4 - The Gap Depth Analysis

The Gap depth determination helps identify where performance gaps emerge. By identifying, a gap between what IS (current) and what SHOULD BE (prediction), one can go through several levels of causes that may contribute or form the gap. Having found the gap location it will be easy to find appropriate solutions. In other words, we have a process (mechanical, organic, information or a business/industrial organization) that has been designed to work within precise specifications. Inevitably, qualitative and performance gaps emerge in a changing environment. Scientists and engineers can predict most of these gaps and can easily compensate through various contingent plans.

Gap #1 is the minimal performance standards gap. Here the question is whether operating conditions meet minimal performance standards: before and while doing. These initial values include quality assurance, resources, logistics, and competencies; and 'while doing' include: standards, rates, infrastructural change (5 levels), quality, cost, in-time (minus) waste. Here conditions either meet or fail minimum performance standards. In the 'scientific' realm, (see right column on Figure 18) this first level establishes and ensures that empirical data collection complies with quality – standardization and documentation of empirical data (see 1st process in OUSM).

Figure 18 - Gap Depth Determination

PERFORMANCE	Begin with actuals – what **IS**. Gap appears as a **SHOULD**. Gap's cause is identified at a *Depth Level*(s). OUSM helps resolve gaps. { **SHOULD BE** / *GAP* / What **IS** }	SCIENTIFIC SOLVING
Level of Causes	Depth Level	OUSM method
1	Minimum performance standards met. Before & while doing	Data collection – 1st Standards met
2	Work expectations met - while and after doing	Quality documentation – verifiable, tools, standards
3	Are methods appropriate? Why/ what is actually being done?	Comparative; ▪ Global ▪ Id gaps ▪ Hypothesis
4	Conducive environment ▪ Research ▪ DMPS ▪ Objective	Causal alternatives: ▪ Strategy ▪ Culture ▪ Implementation ▪ Test ▪ Theory
5	Organization's Purpose, Operation and Style	3DMM; ▪ Origins ▪ How

If minimum standards are met (#1) then the gap may appear at the second level (#2). The second level gap relates to work expectations – the manner or means of doing the work may cause unnecessary delay or expense. In the 'Scientific' column, #2 compares this data identified in multiple areas: multifaceted quality of data/information/ knowledge documentation, timing, value and work expectation. Such comparative analysis can identify information, procedural, prediction gaps. Procedural and predictive gaps may identify the introduction of premature ideological filters, as in the example where in Mongolia scientists excavated a fossilized mother dinosaur brooding upon 22 fossilized eggs. These eggs still contained protein – a surprising condition considering the chemical instability of proteins (Ariel A. Roth, 'Origins,' Review and Herald Publishing Association, Hagertown, Maryland, 1998, pages 242-243). Such a discovery is 'surprising' only because the uniformitarian filter pre-maturely 'forced' millions of years upon such fossils. At this stage in the 'Gap Depth Determination' (Figure 18) this questionable timescale violates #1 and #2 on the '12 Limitations of Science' (Figure 14).

The third level gap relates to methodology – why and what is actually being done? The answer lies in improving the process and product. In the 'scientific' column, the 3rd level gap focuses on selecting causal alternatives. This includes strategic and cultural (management). This should provide the means for proper testing, implementation to ensure conditional certainty.

The 4th gap relates to whether the environment is favorable to scientific research – attainment of conditional certainty. Related to the 3DMM, the environment should allow for proper research processes, implementation of productive policies, and exercise of DMPS and the guarantee of achievable objectives. Under these proper circumstances, the fourth level gap disappears because all conditions are met. The issue of origins and the reconstruction of pre-/historical events can substantially be re-enforced not only with the products of 1st, 2nd and 3rd knowledge, but also with the 3DMM's 27 management plans, infrastructures and other inherent features of the 3DMM features and environment.

3.5 - Rules & Regulations – Supervisory Level Plan (D6)
In business and project management applications

Engineering project management does <u>not</u> define 'Rules and Regulations' in terms of suggestions, recommendations or options. In 3DMM (Fig. 5), 'Rules and Regulations' (R&R) (D6) include a range from short commands that help enforce procedures (D5) to policies (D4). R&R conform to executive plans (D1 thru D3), and align themselves with columnar plans – strategy (D3) and productivity (D9). Vertically, R&R aligns with the Implementation (O6) and Habit (S6) plans.

Rules can come in the form of: simple one-word commands (do not, do, give, carry, wear or turn); phrases (green tags, customer delight); sentences (wear badge/ goggles/ gloves). Regulations usually appear in the form of long directives such as those used in customs and port of entry. Other R&R help modify negative behavior patterns. R&R can be codes that establish 'borders' to activities and scope.

To briefly review: where Policies (D4), as described earlier, reflect the widest discretion to change (guidelines to thinking). And Procedures (D5) provide examples of best practice, programming and flowcharting that need to be updated to ensure that they remain 'living documents' in tune with the Executive plans and Policy (D4); **then** Rules and Regulations (D6) should be viewed a having practically no room for change, thinking or flexibility. R&R in engineering include checklists, directions, standards, codes, tests. R&R is the foundation for data, information, instructions and knowledge that eventually work to define the organization's experience – in whatever set forms and designs. Inevitably, such data-information reflects a legal character that affects or maintains the organization's safety, security, quality, and legal definition and reputation[xlii].

Functional plans (D7 thru D9) track and direct the conversion of inputs into outputs (Product – D9), and provide additional measurable feedback information used to ensure quality and performance. Within the Functional plans(D7 – D9), the R&R (D6) plan monitors quality assurance of input, process and output as specified within the Supervisory management plans D4, D5 and D6. Quality Assessment data (history) is re-assessed to adjust performance. This contributes to development and change within specific ranges and scopes. As examined in Figure 16, change appears at five infrastructural levels of complexity: copies (no change), adaptation (re-alignment), improvement (re-engineering); innovation (re-engineering); invention (hypothesis of the hypothesis

Economics (D8) is the management accounting, auditing, cost control, risk analysis, investment, expenditure, revenues and profits.

Productivity (D9) is the output that must meet the objectives and customer satisfaction. It involves logistics, efficiency and effectiveness testing, reporting (feedback system), elimination of 'noise' and delays.

The Management Design (D1 to D9) level is further refined, tested, calibrated and automated through the columnar screening: directive, processive and info-base functions. For example: the Purpose (D1), Policy (D4) and Research & Development (D7) plans have a directive quality that is designed to provide and set the scope of the 'marching orders' for the following activities in the 'processive' plans: objectives (D2), procedures (D5) and economics (D6 and similarly for the info-base column

The Management Design infrastructure (D1-D9) when properly automated is used to facilitate the management of the Operational Plans (O1-O9) and in turn, the Stylistic Plans (S1-S9). Similarly, the upper infrastructure's plans help define the plans at the design infrastructure.

Rules & Regulations – Supervisory Level Plan (D6) - *Other Applications*

Rules and Regulations are directly linked with quality control (e.g., compliance with standards), authorizations (executive level permissions), and approvals/rejection (specific supervisory level policy and procedure compliance). Where Executive laws have a permanent character, the Supervisory rules and regulations may adjust themselves to the environment – as policies and procedures, while remaining within the scope of the Executive framework (purpose, objectives and strategies). This is evident in any organization, psychology and nature of life.

As a knowledgebase (3DMM), the Christian Bible contains sections with evident *Rules and Regulations* derived from the Executive framework, and reflects Policies and Procedures. One can see these through the application of economics, engineering, construction, and ethics, for example on issues pertaining to chastity before marriage, monogamy, family rules, role of parents, the function of a patriarch, on how to raise children who are destined to become quality citizens in and manage areas of the Kingdom of God. Here are rules and procedures on how to become managers worthy of becoming kings and queens. There are Supervisory rules on building, maintaining relationships with neighbors, how to manage business (payments, scales, borrowing-lending, interest/usury), property, agriculture (annual, seven-year to semi-century cycles, crop rotation, value of crops, trees, etc.), husbandry (selective breeding), accidents, priesthood functions, health, nutrition and many others.

Biological life clearly exhibits rules at their genetic code level. This life is programmed to search for specific nutrition, sense interpretation, and digestive system, muscular, wholesale design to function within the given environment. The functioning eye processes digitally every second as much information as a CRAY computer processes in a 100 hours. Programmed and coded are also many decision-making and strategic functions - operational procedures and instinct. Human Habits on the stylistic management levels (S6) appear to function like programmed R&R (D6) in many living creatures.

Scientific projects and research consist of many R&R. R&R are evident and direct national scientific policy. It therefore becomes easy to identify the strengths and the weaknesses of a national policy, and inevitably, trace this to the scientific executive Purpose.

For example, one needs to visit websites such as the 'Science, Evolution and Creationism: National Academy of Sciences' by the National Academy of Science (NAS).[xliii] Then follow how users implement NAS rules & regulation for scientific research and interpretation - for example the 'True Origins' group;[xliv] and in the many articles that appear in worldwide distribution outlets like Wikipedia.[xlv] Many other disciplines reflect the Rules and Regulations specified in these three sources. These Rules appear in all media - on Discover Channel and National Geographic, National museums, news media, US Federal Courts that are bent on uniformitarianism, and school textbooks at all levels. These guidelines (policies) and rules are evident during high-level science debates.

3.6 - National Academy of Sciences' 24 Rules

National Academy of Sciences' guidelines also provide a litmus test for two types of scientists. The first is the professional scientist who loves his research and goes where research leads. This scientist minimizes ideological influences and may go as far as he/she can to correct ideologically polarized science. The second is the professional scientist who loves ideology more than his research and goes to great lengths to color science and scientific results. The original scientific method helps identify and objectively convert uncertainty into conditional certainty. The second scientist employs tools that bring about issues of demarcation lines between science and religion (specifically Christianity) at the earliest stage of scientific research – the gathering and documentation of empirical data; or maintains scientific research at the 'uncertainty' level in areas where evidence clearly shows that his/her predictions do not support the assumptions.

The *National Academy of Science* (NAS) provides rules and regulations that a scientist or academia is to follow on the path to true uniformitarian science. Through this process it becomes evident that the NAS and its kindred National and State organizations function as a veritable *Scientificist Magisterium.*

The *'Science, Evolution and Creationism'* is NAS's foundational document where a host of participating members from accredited universities, institutes and colleges endorse this document. Here we have: 15 members on the revising committee (bibliographies at the back of the 74-page booklet); 4 staff members; 2 consultants; the document is acknowledged by a 30 member group of presidents, vice-presidents, deans, directors, chiefs, professors, and scholars.

The Preface of this document 'Science, Evolution and Creationism: A View from the *National Academy of Sciences* [xlvi] suggests the following 24 rules for gaining the true uniformitarian perception of what distinguishes true science from pseudo-science (i.e., Creationism and Intelligent Design). All evolutionists use this 'Rules Book.' This includes: science professional, academic administrators, textbook publishers, teachers, government officials, State and Federal Courts, the media, debaters, most Church officials who subscribe to the Theistic Evolution interpretation, and critics/debunkers. All of these express their adherence by word or deed to NAS's authority and their politically correct position. NASs document is equivalent to the 'Communist Manifesto' and Mao's 'Red Book' for the progressive uniformitarians.

NAS's 24 rules can be grouped into six topics:

Group	Rules	Topic
1	1-7	Strategic and tactical use of terminology, concepts, providing new definitions for everyday words
2	8-10	Interpretation of and use of historical data
3	11-14	Biology and the lab
4	15-16	Science's framework for progressive Religions
5	17-23	Challenges to Creation Science and Intelligent Design
6	24	Summary

Rule 1: Use syllogisms: (NASs booklet 'Preface': first paragraph)

a) Suggest a truism – e.g., 'science and technological advances have had profound effects on human life.'
b) Provide a negative contrastive example: 'In the 19th century, most families could expect to lose one or more children to disease.'
c) Make a deduction: 'today, in the United States and other developed countries; the death of a child from disease is uncommon;'
d) Make a lengthy conclusion: 'Every day we rely on technologies made possible through the application of scientific knowledge and processes.'
e) Add four to five additional sentences about the benefits of science: rapid travel, advanced medicines, and higher living standards, orbiting the Earth and trips to the Moon.

Not mentioned here but developed below, is that 'science and technological progress' is to become synonymous with 'uniformitarianism'. That without the uniformitarian perspective everything will remain at a pre-scientific/technology stage of development and at cultism.

Rule 2: Embed redefined terminology and markers. Upon the first or second reading, embedded markers would go un-noticed. For example, the ruling for the use terms and concepts like: 1) 'application of scientific knowledge and processes,' which should read as 'uniformitarian materialism and reductionist processes.' 2) 'Insights obtained from scientific research;' should read as 'insights obtained from the simple-to-complex and primitive-to-modern.' As described above, the uniformitarian Filter pre-determines all phases of the DMPS. This uniformitarian 'acceptable science' predetermines NAS's sentence: 3) '[Science] has given us new ways of thinking about ourselves and the universe;' and offers the following uniformitarian interpretation: '[uniformitarianism] has given us new ways of thinking about our primate origins, and the Big Bang universe.'

Should some readers doubt the 'should read' interpretations in the above paragraph; the following nine paragraphs in NAS's booklet will help remove all doubt. It is here that the hopeful modern scientists receive truth and revelation on uniformitarian 'science' – its conception and growth process. Participating members from accredited universities and institutions endorse the NAS's science booklet. These authorities will not consider anyone or anything to be scientific unless that individual or institution complies with these rules.

Rule 3: The reader does not find in the NAS booklet, that the cornerstone of modern science is the 3DMM (knowledgebase, standards) and its DMPS methodology, which helps convert uncertainty to conditional certainty. Instead, NAS focuses our attention on one single branch of science – biology, which disguised in some esoteric terminology, becomes 'evolutionary biology.' Why not simply write that it is the 'uniformitarian approach to the subject of biology?' (NAS, 'Preface' – paragraph 2)

Rule 4 – NAS states that the prime focus of science is 'evolutionary biology' but then extends, through an inductive leap, this 'evolutionary biological ' approach to ALL the sciences. It seems then that all branches of science derive from the evolutionary biological model? Then NAS suggests that evolutionary biology, together with all science branches contribute 'to human well-being.' This cocktail of sciences contributes to the prevention and treatment of human disease, development of new agricultural products, creation of industrial innovations, the study of new life forms, the relatedness and diversity of present-day organisms; rapid advances made in life sciences and medicine. However, at no time does NAS clarify that the doctrine of 'uniformitarianism' is the foundation to both 'evolutionary biology,' as well as, all of the scientific branches. Furthermore, uniformitarian/evolutionary biology must:

1) Appropriate the works of professional scientists who are not necessarily evolutionists – such as Greog J. Mendel,[xlvii] whose works the Evolutionists borrowed to construct their '*Evolutionary Synthesis*' (1930's) and the '*Modern Evolutionary Synthesis*' (1990's). During this process, evolutionists leave the impression that Mendel's genetic work on inherited traits, exclusively and uniquely supports the Theory of Evolution. However, the Creation and ID scientists also recognize the non-evolutionary interpretation of Greog J. Mendel's contribution to science.

2) Prematurely give an evolutionary spin on new paleontological, biological or astronomical discoveries even though soon after, in countless cases, evolutionists must retract or remove this evidence for evolution when secondary interpretations or new evidence contradict evolutionary predictions[xlviii]

3) Ignore evidence that contradicts the uniformitarian mechanical, organic models of thinking. Such evidence may appear in areas of information and nano robotic technology at sub-chromosomal levels; fossils of modern man discovered at sedimentary layers below those of the Java man; and among many others, the numerous 'out of place artifacts,' and their negation of pre-Ice Age human civilizations.

Rule 5 - Place a disclaimer: for example, NAS provides an example of such a disclaimer on page XI of its booklet: '*Of course, as with any active area of science, many fascinating questions remain, and this booklet highlights some of the active research that is currently under way that addresses questions about evolution.*' This disclaimer states that 1) '*evolutionary*' science does not have all the answers [although answers may exist within OUSM or within Creation and I.D.'s systematic scientific studies]; 2) 'there remain many fascinating questions'; and 3) 'active research provides evolutionary answers'. Such formulation clearly unveils an ideological (uniformitarian)-based faith statement. Here we enter into the realm of optimistic hope, faith and dreaming – in an attempt to help fill the 'scientific' gap.

Rule 6 –Identify a few questions that opponents raise –e.g., 'complexity of life', proof for 'common descent'. Because the Theory of Evolution is at best an organic model of thinking established upon reductionist materialistic tenets, NASs Evolutionists have devised a four-layered red herring: first, evolutionists must associate 'complexity' and 'common descent' questions not with scientific but with religious motives and beliefs. Second, NAS' Rule book states that the uniformitarian scientist should remind everyone that school science should only point to 'scientific' – i.e., uniformitarian approaches and explanations – even though the uniformitarian organic thinking model can't account for 'complexity of life' or 'common descent'. Third, if this sidetracking technique does not work, NAS recommends involving evolution-oriented high profiled individuals (school board members, science teachers, education leaders, policy makers and legal scholars) at the decision-making points, meetings and get-togethers. Such settings should also include: highly graded school and college students, adults who wish to learn about the evolutionary 'fact' and 'process' that account for the diversity of life on Earth. This rule does not mention that such a setting is designed to promote not specifically science but uniformitarian ideological views that may exclude scientific inquiry. The aim is to bend the mind in a way that it will make it function along an uniformitarian ideological track.

If members of the audience persist on addressing complexity of life issues, NAS Consultants recommend a fourth approach - the scientist, teacher or museum guide must link these questions with latten religion orientations (specifically Christian). The true uniformitarian must emphasize that any Christian-based science, such as that implied by the 'Young Early Creation,' must only be viewed in terms of being pseudo-science, and **not** science. Similarly, in such a pre-cast scenario, it is unlawful to mention OUSM, DMPS, 5 Level of Change and the 3DMM. Such an uniformitarian staged strategy points to the implementation of an ideological rather than a scientific content within uniformitarian 'science', communication and motives.

3) Uniformitarians have gone to great lengths in their attempt to prove that 'evolution' is 'fact.' Yet, this reasoning works only within the uniformitarian redefined world of mirrors. As mentioned earlier, a fact is contingent on its framework. People have built multi-century world empires on the 'fact' that the Earth is flat. My current book and Table 7 demonstrate that uniformitarians thread the same path as Flat Earth believers.

If the relentless public continues to seek qualified scientific answers to 'complexity', NAS proposes a *fifth* strategy – that uniformitarian scientists and users apply the *legal option*. They should consult 'legal scholars.' These legalists, together with documented uniformitarian court decisions, have clearly demarcated science (evolution) from pseudo-science (religion and cultism).

Rule 7 – An uniformitarian 'fact' becomes 'fact' when:

1) The consensus of the scientific community accept a fact being a fact ('*Scientificist Magisterium*')

2) The 'nature of science' (i.e., uniformitarian materialism) is contrasted to religion (subjectivism)

3) When enough evidence emerges [*hope*], this cumulative evidence may convince enough religions to conform to the uniformitarian definition of what constitutes to be reality and scientific 'facts' - note the similarity with item #1 – science by consensus. The implication is that when 9/10 of the Christian groups accept the uniformitarian position and a significant minority does not, then the exclusive minority will most likely exhibit heretical motives or delusionary persuasions.

Uniformitarians insist that the concept of 'fact' is somehow equivalent with 'unquestionable reality'. Thus, the NAS booklet identifies Evolution with/being 'fact.' What is overlooked is that such defined 'facts' work within the uniformitarian world of mirrors. As mentioned above, a fact is contingent on its framework. People have built multi-century world empires on the 'fact' of a flat earth. This book shows that uniformitarians thread the same path. Methodologically speaking, such uniformitarian 'facts' and framework still remain within the realm of 'uncertainty.' 'Uncertainty' is subject to investigative tools (e.g., DMPS) designed to lead to 'conditional certainty.' Yet, uniformitarians 'feel' that their hypothesis and theory is reaching 'almost certainty' and is truth' solely within the framework of the uniformitarian framework, without referencing anything resembling DMPS. As discussed earlier, DMPS demonstrates that the hypothesis (alternatives #4) and the theory (#6 & 7) still reflect 'uncertainty' and as uncertainty, must be subject to investigative procedures (DMPS). Upon closer examination, it becomes easy to see that the uniformitarian position on 'facts' is similar to that of those who adhere to the Flat Earth mode of thinking.

Table 7 - Similarities: Uniformitarian and Flat Earth Views

#	Uniformitarian 3-Point Rule	Flat Earth Society's implied 3-point Rule
1	Begin with existing conditions, rates and processes	Begin with existing conditions, rates and processes
2	Extrapolate these through uniformitarian materialism and reductionism (mechanical and organic mental models) to any starting point (origins) – Big Bang, Steady State	Extrapolate these through Euclidian mathematics (e.g., parallel lines) to any starting point (border) – whatever the shape (circular, square, etc.)
3	Draw forward to the present via postulated naturalistic laws using the 'simple to complex' doctrine.	Draw forward to the current status via natural, observable, measureable , geographical laws – simple to complex math
4	Use exclusively a *closed* systems approach. Note: the *open* system approach suggests supernatural participatory forces, and users must avoid this approach at all cost.	Use exclusively the *Euclidian* systems approach. Note: the *non*-Euclidian system approach suggests *abstractitis*, and users must avoid this approach at all cost.
5	Evolution is almost certainty and truth	The flat earth view is almost certainty and truth.
6	Non-evolutionary views are in error, demonstrate latten religious / subjective orientation, and promote deception	Non-Flat Earth views are clearly in error, demonstrate that science promotes distorted views and is fundamentally pseudo-science[xlix]
7	Evolution is science and religion is illusion or pseudo-science	No position

Rule 8 – Scientists must emulate NAS's interpretation of history. For example, NAS writes that 'The National Academy of Sciences has had a mandate from Congress since 1863 to advise the federal government on issues of science and technology…' Note that during the American civil war (1863) the American government did not mandate an *uniformitarian interpretation of science*. It is necessary to research NAS' history to identify under what circumstances NAS installed this un-constitutional uniformitarian *ideology* (see Table 8 – 'Comparison: Religion and Ideology'). This NAS ideology has removed geometric natural law-based Constitutional foundations and installed itself as the official State ideology – in violation of Establishment Clause.

Rule 9 – Scientists must provide historical re-enforcement whenever they communicate evolutionary topics. This means that scientists must list development and updates in paleontology. Such updates are to create the aura of 'compelling evidence about evolutionary history; understanding about molecules, DNA sequences of humans, relationships among species. Where the fossil record is incomplete [NAS introduces a new field] 'evolutionary developmental biology' which should show how genetic changes...should fill the gap'.[1]

This general and synthetic history of uniformitarian developmental 'evidence' brings another rule:

Rule 10 – Mention historical legal court events and decisions (see Rule 6). When this does not quell down issues of 'controversy' then uniformitarians must simply resort to 'scientific consensuses,' how the 'scientific community' views such 'scientific' issues. Today, such collective bodies rubberstamp all uniformitarian positions. These and others such bodies form what I call the 'Scientificist Magisterium.'

Rule 11 NASs booklet entitles Chapter 1 as '*Evolution and the Nature of Science*' and subtitles it: '*the scientific evidence supporting biological evolution continues to grow at a rapid pace.*' Here uniformitarians/evolutionists follow directions to describe Evolutionary Mythology. The rules are:

1   Identify a stratum from which the scientist will bring out an 'intermediate' fossil
2   Date this fossil by allocating it several hundreds of millions of years (use the geologic column) – Note: implied - never introduce dates that are less than 6000 years.
3   Pepper the descriptive narrative that describes the form with many 'maybes', 'probably,' 'it is believed', etc.
4   Use the services of science artists to help create a full-blooded creature that has moved, breathed, functioned in an unusual and exotic environment
5   This improvised creature finds and obtains its nourishment from an environment of many other co-existing and competitive creatures

In other words, create a full-scale Hollywood-like production. For example, the fraudulent 'Nebraska Man' had been provided with fictitious bones, flesh, and clothing and with a family – all fleshed out from the uniformitarian interpretation of a pig's tooth. This is just one example among many.

Rule 12 – Chapter one's sub-section is also subtitled as '*Biological evolution is the central organizing principle of modern biology.*' This means that DNA mutates through natural selection, as groups of organisms are isolated and overcome challenging environmental conditions over multiple generations. This oversimplified linear explanation demonstrates ideological rather than scientific evidence. Evolutionary scientists are limited to consider only linear change and cannot speculate on how macroevolution occurs. Yet, issues of chance cannot account for concepts of information creation, duplication, conversion, self-maintenance, self-correction, self-replication, and self-deletion through change levels requiring re-engineering, re-design and invention, as well as, concept of supervisory, executive and infrastructural systems. These are required non-uniformitarian explanations – e.g., information and nanobot technology and geometric natural law.

Rule 13 – NAS urges scientists who work in uniformitarian labs also to draw upon some industrial examples. Here scientists are to use principles of natural selection to develop new molecules that have specific functions (page 9 on NAS booklet). The 'industrial lab' is an area where natural selection is challenged most. The uniformitarian author brings industrial organic examples that would suggest organic adaptation or 'chemical evolution.' Yet neither the uniformitarian scientist nor the NAS booklet mention anything about pre-requisites for such adaptations or chemical evolution - information and nano-robotic functions at the sub-chromosomal level –that are key requirement for change in genetic structures.

Rule 14 – Another sub-section of Chapter 1 is entitled: '*Scientists seek explanations of natural phenomena based on empirical evidence.*' This is another opportunity for the evolutionist to make a clear distinction between empirical sciences vs. information coming from outside the natural closed system (non-materialistic influences - religion).

NAS mentions nothing about geometric natural law, legal-based history, 3-D management models, or DMPS. Although NAS talks about testable evidence, reproducible experiments, predictions and refutability (falsifiability), nothing has addressed the pseudo-scientific nature of the uniformitarian process (see the Score Cards in this book). Furthermore, in this sub-section NAS drills again about the 'fact' of Evolution- see 'facts' addressed above in this book.

Rule 15 – The scientist is supposed to show that '*acceptance of the evidence for evolution can be compatible with religious faith.*' NAS presents here the various Hybrid Uniformitarian religious views. With this, the NAS booklet includes full page-quotations and 'excerpts of Statement by [Uniformitarian] Scientists,' who confirm the feasibility of such marriage. Such a marriage of convenience is possible as long as such religions fall within uniformitarian definitions of what constitutes religion (see Rule 22). At the same time, it becomes evident that religions that acquire 'uniformitarian realism' and by definition metamorphose into, or be pantheistic.

Rule 16 – Most evolutionists claim that the theory of evolution applies strictly to the biological domain. Evolutionists take every opportunity to ridicule skeptical opponents who apparently cannot understand this simple reality. Yet, after having focused our sights on this biological domain, NAS's Chapter 2 suddenly addresses '*The origin of the universe, our galaxy, and our solar system produced the conditions necessary for the evolution of life on Earth.*' Are these evolutionary biological subjects? In this Chapter 2 NAS discusses radiometric dating, living things appearing in the first billion years of Earth's history, how fossil records help document the occurrence of evolution, common descent, comparison and similarities between the chimp and human DNA. NSA consultants go out of the way to talk about fossil footprints in a manner that would help dispel or deflect possible interpretations of dinosaurs and human footprints appearing in the same sedimentary bed.

What evolutionists fail to note is that when they expand beyond the 'biological' sciences, they inevitably identify what lies at the foundation of the evolutionary biological science – it is the common denominator - the *Uniformitarian ideology* (see Tables 6 & 7 and 9). It is within this framework that all biological and non-biological scientific topics are treated.

Rule 17 – NAS's Chapter 3 addresses the Creationist and Intelligent Design Perspectives. Under the sub-title, '*Creationist views reject scientific findings and methods*' the uniformitarian rule is to place the Creation and Intelligent Design among 'religious' groups but in a separate category from the Hybrid Uniformitarians (Theistic Evolution, Progressive Evolution, and Old Earth Creationists).

The reader must however be also aware that NAS's consultants believe that '*no scientific evidence supports [hybrid uniformitarian] viewpoints*' (see NAS booklet page 38). This verdict, therefore, in-spite of earlier uniformitarian positive pronouncements towards Hybrid Uniformitarianism and other world religions, still places these Hybrid Uniformitarians on the eighth or seventh ring of the uniformitarian hell (imagery taken Dante's poem entitled 'Hell'). Materialistic Uniformitarianists still perceive the Hybrids as being cultists, just as they see the Creationists (two millennial Christendom), the Young Earth Creation scientists and the I.D. scientists who in the uniformitarian hell reside on the lowest – ninth ring.

Rule 18 – Uniformitarian scientists cautioned those who reject the billions of year-old ages. The Earth is canonized to have an age of 4.5 billion, while the Universe's age is set at about 14 billion years. Rejecting these doctrines is virtually to reject '*not just biological evolution but also [the] fundamental discoveries of modern physics, chemistry, astrophysics, and geology*' (see NASs booklet, page 38). Evolutionists must. therefore wholeheartedly reject the worldwide Flood, a recent special creation, geometric natural law among other concepts. This is because evolutionists disagree on how sedimentation processes occurs, and that sedimentary deposits, under such circumstances, must be evident on top of some of the Earth's highest mountains (??).

This rule 18, presents science in terms of axioms. However, these axioms reveal several unscientific positions:

1) Uniformitarians claim or define science as being synonymous with the uniformitarian materialist and reductionist ideology. Yet, it is clear that science existed prior to the establishment of the 19th century uniformitarian ideology.

2) The scientific method (DMPS) helps revolve 'uncertainty', where at the fourth stage, scientists describe, quantify, examine, test, apply, and predict the evidence within all possible alternatives – from evolutionary to creationist. To exclude all scientific options (alternatives) is to do so on ideological grounds which renders the scientific method obsolete (see Figures 9 thru 13)

3) True science must address the '12 Limitations of Science,' which also includes the objective unfettered scientific method (OUSM). Uniformitarians violate all of these scientific tools while Creationism and Intelligent Design do not

4) If NAS Science Consultants had done their homework, they would have realized that the Creation scientists have long ago addressed sedimentation and the sedimentation on top of mountains and continue to address this subjects scientifically today (see www.icr.org). For example, the mountain ranges that we seen today have been uplifted through the hydro-tectonic processes during and after the global catastrophic events – third singularity. As described above, the CRAY computer software had helped provide evidence for the feasibility of these events. The Los Alamos Laboratories provides supporting evidence for global catastrophic events and predictions.

Rule 19 – NAS's rules address the fossil record, which is a *'rich and extremely detailed record of evolutionary history that paleontologists and other biologists have constructed over the past two centuries and are continuing to construct.'* Yet, NAS forgets to mention that this view does not agree with research results and publications offered under their own (uniformitarian) auspices. For example, the fossil record is the single area, which disproves evolutions. For example, Goldschmidt proposed to explain vertical macro-change by suggesting an event of instantaneous-speciation, saltation, or systemic mutation that produced new groups – now known as the Hopeful Monster theory.[li] Niles Eldredge and Stephen Jay Gould,[lii] based on Ernst Mayr's[liii] theory of geographical speciation, published a work in 1972 where both showed a greater emphasis on the development of a stasis theory – i.e., 'punctuated equilibrium.' This punctuation was to help explain dramatic changes when phenotypic evolution occurs in rarely localized conditions, coupled with rapid events of branching speciation (cladogenesis).

NAS Consultants and founders of the 'Modern Evolutionary Synthesis' axiomatically 'reject' both of these views and do not falsify them scientifically. What was the reason for discarding such scientific conclusions? The answer is that these views did not match the uniformitarian doctrine and predictions. Yet, the 'Young Earth Creation' scientists have a viable scientific model that predicts and explains these paleontological gaps. Again, when new techniques, such as computed axial tomography,[liv] are used to learn about the internal structures and composition of delicate bones of fossils, and the Tiktaalik fossil, the uniformitarians, specifically in this case, NAS substitute science with uniformitarian fiction. By themselves neither the computing axial tomography, nor the fossil remains of Tiktaalik fill the fossil gaps. Nor do either explain the 'scientific' aspects of evolution theory.

Rule 20 – NAS wants its proponents to challenge the non-uniformitarians with statements or questions such as '*Nowhere on Earth are fossils from dinosaurs, which went extinct 65 million years ago, found together with fossils from humans, who evolved in just the last few million years.*' This statement is a variation of the common argument '*I dare you to find a human footprint in the Cambrian rock.*' In other words, both issues must be argued from the Uniformitarian ages approach!

First, if such evidence (dinosaurs-man contemporary existence, or human footprints in Cambrian rock), were to be uncovered, one would first have to examine NASs uniformitarian past reasoning, methods and Rules, and the scientific manner in which Evolutionists handle 'gaps in the fossil record' (see Rule 18). What would be NAS / uniformitarian reaction if such discoveries were to be made and documented? Would NAS follow OUSM - the 'objective and unfettered scientific method'? What 'scientific' process would the uniformitarian scientists and its Magisterium follow to further illuminate this 'out-of-place' scientific evidence and bring it into scientific light? The risks are high that such evidence would never see the light of day. Yet, there had been scientific evidence before the highly financed and 'accredited' institutions had a chance to lay their authorizing hands on it. Such empirical evidence had been documented in scientific journals! Such as the clear imprint of a child's moccasin footprint in what is considered to be Cambrian rock, and many other cases summarized in Appendix 2.

Rule 21 – NAS writes in its booklet, that Creationists state that the '*theory of evolution must remain hypothetical because there aren't any ways to test it scientifically*'. In reply, NAS restates the *direct* or *empirical* scientific process, and then suggests that many scientific 'facts' can only be *inferred* indirectly from existing empirical evidence. For example, NAS reveals that evolution is apparently part of the *indirect* or *inferred* body of knowledge/method. NAS recommends hypothesizing, but forgets to mention that the hypothesis remains within the realm of 'uncertainty.' This uncertainty in scientific terms must pass through the DMPS/scientific phases, instead of remaining an ideologically viewed 'almost certainty, and truth.'

At the same time, NAS perceives that the same hypothesizing in the Creation and Intelligent Design scientists' hands becomes religion. It is also precisely at this hypothetical level that Evolutionists are the most vulnerable and at this point where Creation and I.D. scientists disagree with uniformitarian interpretations, assumptions and conclusions. Hypothesis is where Creation and ID scientists reveal their religion, while uniformitarians reveal their ideology.

NSA alleges that Creation and Intelligent Design scientists ignore much of the scientific 'facts' that evolutionists bring forth. This statement clearly shows that evolutionists are not well read on Creation scientific literature. It is the Creation and Intelligent Design scientists who know and follow quite closely all scientific research done by evolutionists in this the USA and around the world. What these Evolutionists mean by heir accusation is that the Creation and Intelligent Design scientists do not necessarily accept uniformitarian (materialist and reductionist) interpretations, evidence and predictions.

For example, during the many high level Evolution-Creation university debates held during the late 1970's and early 1980's; the Creation scientist (www.icr.org) had held a significant advantage over their Evolution counterparts because in a major part during the debates, the Creation debaters quoted the scientific publications that appeared in evolution publishing establishments. It is clear that hundreds of thousands of 'Evolution' scientists are honest, love their investigative scientific work, and follow where the evidence leads them. They publish their data and results, which Creation scientists study, use in their 'Creation Scientific Models' and Evolution Scientific Models, and present the result during the debates. Through this, the Creation scientists had been able to provide better scientific explanations and prediction than those produced by Evolution debaters.

On the other hand, in view of what NAS wrote in many parts of its booklet, it has becomes clear that its own scientific experts and consultants are:

a) Not aware of what the 'Young Earth Creation' scientists publish[lv]

b) NAS scientific experts and consultant intentionally ignore Creation scientists' publications and mention of the Creation Scientific Model

c) Use the NAS media to misrepresent Creation scientists' positions in order to dissuade the public from even considering reading Creationist scientific literature.

Rule 22 – NAS makes questionable allegations of what scientific data Creationists accept or reject. Based on questionable evaluations, NAS science consultants constantly jump to premature conclusions. For example, NAS writes *'Creationists reject such scientific facts in part because they do not accept evidence drawn from natural processes that they consider to be at odds with the Bible. But science cannot test supernatural possibilities.'* As observed in Rule 21, such a statement clearly shows that NAS' science consultants and authorities have never read or understood the 'Creation Scientific Model' and the results that the model produces. NAS want the public to conform to the erroneous conclusion that Creation scientists do not recognize scientific natural processes due to their Biblical tunnel vision. Actually, Creation scientists publish ('Acts and Facts' see www.icr.org) comparisons of data processed and interpreted through both the *Evolution Scientific Model* and the *Creation Scientific Model*.

At this time, it is interesting to recognize how uniformitarians determine what is supernatural and what is not. Based on 100 to 500 year-old documentation, here is the uniformitarian procedure for determining what constitutes to be supernatural and what is not:

1   Formulate an uniformitarian definition of what is supposed to constitute theism and theistic thinking[lvi] The definition provides an open system that allows supernatural participation. The Uniformitarian ignores geometric natural law.
2   Uniformitarians affirm their closed system view – materialism (closed system)
3   Based on #1 and #2 the uniformitarian counteracts to its own view of 'irrational theism.'
4   Contrasts 'irrational theism' #1 with uniformitarian science #2[lvii]
5   Compare #1 with #4 and conclude that #1 is absurd.

The syllogistic process for this is something like this:

1   U says that X=T
2   U says that U=M
3   U = M≠T
4   X≤≥M
5   X≤≥M = U(M) [-X(T)]

Being a rationalist/empiricist approach to defining divinity (X), inevitably leads to providing a *pantheistic* definition of divinity. As described above, within the 3D Management Model, this formulation is located at the Supervisory Policy (D4) row at the Management Design level and not at the Executive Purpose (law, identity, scope) row where the original Christian purpose is established. As such, the Supervisory policy level can only formulate issues from within time, space and resource content, and not from the geometric natural law perspective.

Briefly: the rationalist/empiricist approach begins with B (uniformitarianist) defining what X /T are, inevitably perceiving that X is actually T, which is a wrong assumption. This assumption is already pre-filtered for an X perspective (pantheism) i.e., the Supervisory (D4). There is no reference to a 3DMM nor DMPS, which define the objective criteria for 'conditional certainty.' In other words, the supervisory approach is a subjective, relative and algebraic approach - syllogistic argument. There is a missing mechanism for converting 'uncertainty' to 'conditional certainty' (DMPS). Because of the missing DMPS, the pantheist does not refer to the DMPS historical-base, standards, criteria, codes. DMPS requires psychological qualifications, the emphasis on the use of only one alternative (#4) (syllogistic, axiomatic and algebraic) – the pantheist disqualifies him/her/itself on this pre-filtered application. This axiomatic position has no objective (lawful) reference point, just an subjective ideological one. Such a twisted approach to DMPS and 3DMM enters a high-risk condition. This subjective ideological approach simply reflects 'naïve realism' – pre-determined uniformitarian Filtration process (#5 in DMPS). This rejects the Executive Purpose (D1) and reduces Management Style (S1 – S9) to Operational management plans. The 4-point uniformitarian realism (UR) is 'naive' because, as shown above, the method attempts to short circuit the DMPS and remains within perpetual 'uncertainty.'

One does not have to go far to help understand uniformitarian realism ('UR'). The Soviet 'Socialist Realism' policy has amply theorized and placed into action a similar option into practice for decades. These uniformitarian Soviet filters affect not only how reality is to be portrayed in the arts (literature, visual, plastic, music), but also applies to education, media, history, and daily conversations. The Socialist Realist directive is to '*Identify in today's world those elements and actions that represent the best examples that will exist in tomorrow's communist world; and, at the same time, downgrade that which will not be seen in the future communist world.*' In this case, those things that will not exist will be: bourgeoisie, capitalist exploitation of the workers, religion/cults – and specifically Christianity and its ethics, private property and rugged individualism among others. The Marxist uniformitarian future shows an international socialism upgraded to world communism, dictatorship of the proletariat and collectivism, among other. This is the dichotomy represented by Socialist Realism.

Uniformitarian critics may point to Creationist pre-conditioned view of a Supernatural Creation that affects the 'scientific' (materialistic) process. History, however, has shown that Christendom had created a favorable environment where the scientific method thrived. This scientific approach recognized and implemented the 3DMM, DMPS and allowed for the quantification, investigation and testing of all Alternatives/Hypothesis (#4) through rigorous procedures. On the other hand, the Uniformitarian approach (Secular, Marxists and Nazi) compromises and narrows this objective approaches at the Filter #5 level. As noted earlier, the secular/pagan Roman Empire could not make Christianity fit into the relativistic Pantheon of Cults-Religions-Ideologies because original Christianity had reflected Geometric Natural Law (see Figures 20 through 27). This same consideration applies to NAS Rule 20. The scope of geometric natural law lies outside the uniformitarian realm since Uniformitarianism functions within algebraic rules - see Figures 19 and 23.

Rule 23 – NAS portrays Intelligent Design scientists as being a group that represents a variation of the Creation model. Actually, ID group had emerged from the Neo-Evolutionary group. These originally Neo-Evolution scientists became dissatisfied with the 19th century pseudo-scientific baggage that the Evolutionary Theory carried – specifically – uniformitarianism. ID scientists with the help of their super-high-tech laboratory equipment were able to penetrate into scientific depths and 'data' 'information' that has never been suspected or even dreamed of. ID, in search of a

Scientific Model, went beyond the mechanistic, organic and processive models of thinking. They explored the information model of thinking. It is within this context that ID scientists considered a new scientific language. This language was unknown prior to the 1980's. These originally neo-evolutionists discovered an unimaginable world at the sub chromosome nano-level. This sub-level of activity within an 'organic model' boggled the human mind. This sub-chromosomal environment has nothing comparable in the modern human world's technology. Here, engineering is so complex, refined, precise reflecting unimaginable management and engineering codes and standards. The honest scientist can only describe this highest complexity by using superlatives, and in areas of causalities ID can only extrapolate an Intelligent Designer concept – a teleological[lviii] view.

At the same time, evolutionists continue their attempt to 'explain' this information-based model in terms of the mechanistical, biological and processive mental models. By analogy, it is like attempting to repair a computer chip with a magnetized Philips screwdriver. Therefore, the best outlet that uniformitarians can have is to close the door on this area of research and accuse ID scientists becoming cultist freaks.

Rule 24 – NAS concludes by re-introducing issues such as:

a)   The application of strict uniformitarian curriculum rules in the public schools
b)   Reminding the public of specific court decisions that ruled in favor of evolution and against religion
c)   Emphasizing the separation of Church and State

NAS writes that '*Learning about evolution is an excellent way to help students understand the nature, processes, and limits of science in addition to concepts about this fundamentally important contribution to scientific knowledge.*' This sentence concludes by emphasizing that only the uniformitarian materialist ideology can be the acceptable and authorized science and worldview. Yet, this ideology is an algebraic rule-based approach that, as developed above, reflects a truncated 3D management model, excludes executive plans, and re-interprets all management style plans that affect ethics, attitude, and culture. Few recognize that this uniformitarian process and its filters are used to continuously brainwash individuals, groups and societies into a reductionist state of mind that prioritizes matter, physiology, behavior models and an economics hierarchy. In other words, every issue that users address must be reduced to purely local and subjective hands-on applications.

Uniformitarians (NAS) have essentially designed a 'debunking' language and strategy to help them sustain this subjective and reduced momentum. This is particularly evident on http://www.trueorigins.org/ where NAS rules put to practice[lix]. Instead of establishing a background or platform for a scientific debate, the NAS guidelines, language, cynical attitude, adversarial orientation ignores the background issues of 3DMM, DBSP, OUSM, etc. In such discussions that are not dialogs, uniformitarians:

a) Function and attempt to place themselves in the role of evaluators and debunkers who use NAS rule book
b) Assume superiority roles and stay on the offensive
c) Evaluate subjectively their opponents' intellectual and ethical competence
d) Demand that their opponents produce sources, references only to dismiss these should these not appear on the Scientificist Authorized list, and present highly mendacious evidence
e) Provide not options to studies or interpretation of data and issues, but criticize the text or ideas line-by-line, offer single worse case scenarios and lose touch with the content or the original objective or purpose for the debate. In such a setting, the uniformitarian opponent is expected to spend most of his/her time correcting and providing alternatives for solutions
f) Preach on the scientific nature of evolution and cultist nature of religion - specifically Christianity. Distinguish between Church and State. Court decisions
g) Read Creation and ID research only through uniformitarian approved reviews
h) Assume all kinds of Creationist and ID assumptions, perspectives and beliefs
i) Never use historically supported trend in the development of science or the scientific method (3DMM, DMPS, 5 Levels of change management, OUSM, etc.) instead under uniformitarian rules (UR, ER), reductionism, closed systems, and naïve realism are ensured. Some honest evolution scientists try to work within these guidelines, but many escape the naïve realism and publish objective research.

After Uniformitarian Realists, exhaust their script, they become functionless – a clearly non-scientific syndrome akin to that exhibited by cultists, activists and fanatics. At the same time, it becomes evident that this uniformitarian fragmented and cynical approach to what is to be a representation of 'scientific' discourse, with an absence of Management Style, points to a hidden agenda, and a resulting decent into barbarism.

NAS again urges its promoters to reintroduce the role of Court decisions – similarly addressed in Rule 6. See additional analysis of Court decision in section 5.1 below. It is worth noting, however, that true American Constitutional law is based upon geometric natural law. Yet, uniformitarians purposefully ignore and bypass this through various intermediary precedents. Evolutionists glorify themselves on their favorable court decisions, but fail to notify the public of its dangerous direction and precedents. The 20th century application of such Uniformitarian-based law and court decisions resulted in justification and implementation of genocide, concentration camps, relativist morality, secular State education, initiatives to eliminate Christendom in uniformitarian Communist and Nazi countries. Uniformitarians single out Christianity because Christianity had historically been the most effective barrier against pagan uniformitarianism – e.g., pagan Roman Empire policy.

These are NASs 24 rules and are the foundation for a new authorized uniformitarian scientific worldview, as well as, State authority. This uniformitarian initiative has penetrated all American institutions. This initiative has also put on notice the original Christian majority of its status as haters and of semi-sane medical patients. Each scientist, publisher, teacher, administrator, court judge, education institutional, curriculum designer, university professor and dean must conform to uniformitarian rules, policies and worldview if they wish to continue in good standing. This uniformitarianism also lies at the foundation of the accreditation process for education and other institutions. Deviators from these rules automatically become pseudo-scientists, religious fanatics, heretics and cultists, who will never have access to uniformitarian research material, opportunities to publish in scientific journals, teach or research in 'accredited' institutions.

It is now easy to understand where debunkers get their confidence and inspiration when they tackle Creation and Intelligent Design scientific works and beliefs. These critics now do not even have to research Creation or Intelligent Design scientific works. Alternatively, if they had done some research, they no doubt would have interpreted it through uniformitarian filters. By wearing uniformitarian goggles, they will almost find it impossible to implement the 'objective and unfettered scientific method' (OUSM). To prove this, one needs to read the difficulty that the ID scientists had to undergo before they set themselves to discover the objective unfettered scientific method.

NAS may be right in one respect in identifying contemporary 'Young Earth Creation' scientists with mainly Protestant Fundamental denominations affiliation. There is even a greater number Hybrid Uniformitarian groups. Here we find also the Anglicans, Roman Catholics and Orthodox Christians.

However, NAS uniformitarians miss the point when they point to Fundamentalist scientists and their 'Young Earth Creations' positions since the six-day recent Creation view is as old as Christendom itself - two millennia and go further to the first day of Creation in the Old Testament. It is this two-millennial Christian view that:

a) Distinguished between Christian vs. pagan worldviews
b) Recognized a recent creation with three historical singularities
c) Documented geometric natural law upon which all of Creation functioned
d) Upheld that man is made in the image of the Eternal Lord God, and not in that 'man made in the image of a primate' as some modern cultist groups maintain.
e) Daily promoted that man daily qualifies for the Kingdom of God
f) Mankind can create, and fill the knowledgebase with accurate knowledge (science) by solving uncertainty through DMPS, discovering geometric natural laws, standards, and through this lead mankind to a better and perfect civilization – a

Kingdom of God on Earth, rather than qualify for and participate in the construction of the Kingdom of Babylon.

g) Maintained favorable conditions where science and the scientific method were developed and applied. Christian scientists and theorists who used the original scientific method had established all original branches of science.

This book examined Uniformitarian Rules (NAS's booklet), and Christian Creation also has scientific management *Rules*. The 'Young Earth Creation' scientists represent these in part. These Creationist Rules are evident in:

Table 6	Comparison: DMPS, Uniformitarian and Creation Scientific Methods
Table 8	Religion and Ideology: Similarities
Figure 14	The 12 Limitations of Science a Scored Comparison
Figure 22	Qualitative Infrastructure & Energy Densities
Figure 24	A Comparison of Geometric Law – Derived or Revealed
Figure 25	Comparison of Filters – Evolution and Creation Science
Figure 26	Genesis Creation: 7-Day Infrastructural GNL Interpretation

CHAPTER 4 - RELIGION EVERYWHERE

For thirty-five years, hundreds of Evolution, Young Earth Creation, Intelligent Design and Hybrid Uniformitarian books, videos, textbooks, and other media produced to justify, confirm and analyze the various approaches to science. Too many Court decisions[ix], however, continue to reflect uniformitarian rather than American Constitutional criteria for defining what constitutes science and religion.

Perhaps two typical journalistic questions may help direct some light on this issue.

1) 'Do you think that teaching evolution goes against religious beliefs?'

2) 'State and Federal courts ruled in favor of Evolution's approach to science rather than that of Intelligent Design, Creation or Hybrid Uniformitarians. How do you interpret this ruling?'

Upon closer examination, the editor's first question contains at least 22 assumptions, inferences and interpretations (see details in Appendix 1). These are also similar to those used by Courts that pronounce verdicts on what constitutes science and quackery.

To examine these questions, it is necessary to define the basic terms and concepts. Among these are:

1 Similarities between ideology and religion
2 Evolution-Creation-I.D.-Hybrid compliance with the 'OUSM'
3 Awareness of the econo-mythological vehicle within society
4 Man's brain prioritizing capacity
5 Algebraic vs. Geometric approaches to constructing reality (science, models, change levels)

4.1 - Ideology's Similarities with Religion

The 22 editorial/court assumptions, inferences and pre-built conclusions clearly reflect an ideological view rather than a scientific one. This is particularly true since these 22-based pre-built conclusions reflect the uniformitarian axiomatic structure.[lxi] Here uniformitarianism:

a) Arbitrarily rejects or negates several 3DMM management plans, infrastructures thus introducing reductionism and relativism
b) Preempts the alternative process (#4) by its filtering values (#5) (see DMPS-Figures 9 thru 13)
c) Narrowing the 'alternatives option' (#4) creates 'naïve realism' (UR/ER/akin to Socialist Realism)
d) Pre-determines the 'validity' or ER results, thus maintains 'uncertainty' instead of allowing the process to reach conditional certainty
e) Does not process hypothesis and theory's 'uncertainty' through the DMPS stages. With this, uniformitarians ascribe an 'almost certainty' to hypothetical and theoretical inherent 'uncertainty' factors.

These deficient factors, which violate scientific methodological procedure, clearly move uniformitarianism from the scientific realm into the ideological realm. The differences between ideology and religion are very narrow.

Table 8 - Comparison: Religion and Ideology [lxii]

RELIGION	IDEOLOGY
Body of beliefs, doctrines	
A set of beliefs concerning the cause, nature, and purpose of the universe, esp. when considered as the creature of a superhuman agency or agencies, usually involving devotional and ritual observances, and often containing a moral code government the conduct of human affairs.	The body of doctrine, myth, belief, etc, that guides an individual, social movement, institution, class or large group. In the philosophical area: it is the study of the nature and origin of ideas. Exercise of theorizing of a visionary or impractical nature.
Social contribution and Adherence	
A specific fundamental set of beliefs and practices generally agree upon by a number of persons or sects. A defined body of persons that adhere to a particular set of beliefs and practices, e.g., World Council of Religions.	This is a body of doctrine, myth, etc., with reference to some political and social plan, as that of Fascism or Communism. Included are the devices for putting it into operation.

In both cases, the beliefs, doctrines and policy move societies and individuals. If religions identify a supernatural component, the same component is evident in ideology – i.e., philosophy, which helps outline ideas, causalities and purposes. In both cases it is where vision, hope and the seemingly impractical is accepted.

However, both religion and ideology are a necessary cohesive component that establishes an 'econo-mythological' vehicle that helps carry society through the challenges of uncertainty (for more detail, see section 5.3 below). Uniformitarianism does not identify this relationship since it rejects 3DMM, DMPS and other factors.

4.2 - Awareness – Econo-Mythological Vehicles

The human mind seeks to confirm its self-awareness within its environment by using various tools. These tools must function effectively, efficiently and competitively. From the limits of what animals can achieve, mankind expands its awareness to multiple areas that help improve survival. With the use of tools, machines, by identifying landmarks, and potential borders, documenting the history of stellar observations, by creating navigational charts that help measure distances (space), identify universals or standards, distinguish among stellar, lunar and local times - mankind continuously has refined a knowledgebase based on standards, codes and mathematics. Mankind has developed scientific instruments that help it confirm, interpret and predict locations, processes, causes and outcomes.

Mankind has been awed by the universe's structure, its mathematicability and what all this has to offer. Mankind had used this structure to set schedules for seasonal and agricultural cycles, to improve productivity and develop precise navigation into unknown and changing environments. It has also designed precise calendars that allowed for the forecasting of realities through which mankind must be prepared to pass.

Today, we may be better equipped to observe stellar, galactic, multiple and mega clusters of galactic phenomena, but we are simply continuing the long tradition that had started at the dawn of time.

This synchronized awareness of the celestial, temporal and economic phenomena has been refined, coordinated and incorporated into practice by the mythological event. This event is a set of procedures at various operational levels. The combination of the mythology-economics infrastructures have become the vehicle used by passengers who travel on a quality assured course.

This vehicle reflects a hierarchy of values that everyone recognizes as 'reality.' This reality allows mankind to identify and manage borders, prevent accidents (risk management) and prosperity. This total production, distribution, construction and discovery have sustained existence. This has occurred everywhere where civilizations emerged.

By tracking the road to the roots of civilizations, we begin to see how the management of mythology has helped coordinate or discoordinate reality. Through a number of means, citizens of any of these civilizations were able to adjust, subdue, force, direct, organize and motivate individuals, groups, social classes, management on any given economic infrastructure. Mythologies, infrastructures and economics helped manage and introduce different types and levels of outcomes. Such outcomes remained stable, static or dynamic platforms of civilization. Alternatively, these outcomes may have served to integrate, galvanize or deteriorate societies. Some means established benevolent as well as exploitative conditions. These civilizations had presented a universe in terms of rational, organizational and eternally discoverable contents, or in terms of accidents.

History has helped identify five social structures (vehicles) that have helped coordinate management, economy, and labor under their corresponding myths:

The first econo-myth is the master-slave model shows a privileged master class ruling over unprivileged masses of labor, slaves, and the vanquished. The slaves' nose-close-to-the-ground myth includes State-devised myths that reflect polytheistic realms. These realms contrast degrees of gods' capacities and the masses' awe of the seemingly visible interplay between the divine and human. Such interplay evokes and over-reaches all aspects of life from conception to the grave, work, play and provides a purpose for existence. We see this expressed in forms of materialistic or mystical polytheism.

The second econo-myth has warlords who maintain a centralized, imposing yet relative and capricious power. This power affects all classes: merchants, tradespersons, owners and serfs/slaves. These classes conform, reflect and entertain adaptive and interpretive behavioral patterns. Their members do not wish to appear to fall outside the task of fulfilling the ruler's will. In such an environment, myths stress material opportunism (atheism and skepticism), and polytheism coupled with mediating mystical powers.

The third econo-myth reflects coalitions of city-states that either have or do not have a single coordinating ruler. Merchant families and trades persons form the core of these city-states. The prevailing polytheistic myths may expand and experiment with a systematic theism that reflect preferred rules of conduct, some underlying principles of economics, some social contract and list of etiquette of convenience.

The fourth econo-myth capitalizes on the expanded and formalized rules, procedures (protocols) and policy-based associations that have helped expand the economy. A mercantile-banking system and think tanks perhaps expanded to meet an imperial requirement and scope. Citizens, merchants consumers conceive some pluralistic myths (e.g., a State sponsored Pantheon of subjective and relativistic cults) that include atheism (pantheism – i.e., nature worship), polytheism and various brands of monotheisms. These myths function in, and help promote a temporary intermediate democratic filtering system. Eventually, overburdened by the 'wishes of the people', and usury's corroding features the now culturally debased and bankrupt civilization settles for an emerging 'federal emergency' authoritarian dictatorship. This new rule, promotes 'just-cause' solutions that eventually function on a lower infrastructural level that may include genocidal de-population schemes. Continuously new common denominators (collectivism, racism, class) emerge to represent the 'more refined' democratic standards (masses).

The fifth econo-myth vehicle reflects the original definition of the republic founded upon the universal and objective principles of geometric natural law (e.g., American Constitution). With its program of continuous improvement, self-perfection and search for scientific and technological breakthrough the republic increases an accurate knowledgebase that ensures that the population and its representative government function within the ever-discoverable geometric laws. This initiative leads towards individual and social self-improvement or self-perfection. Natural laws discovered or revealed through special creation reflected in geometric natural law, covenant law and contractual precedents.

These five econo-mythological vehicles establish and contain four continuous conditions adapted for mankind's own awareness, existence and control:

#1-Matter (cosmos, economics, tool, management and resources)

#2-Interpretation (decision-making and problem-solving, cause, trends, quality, systems, priorities and value)

#3-Degree and nature of change (static, life cycles, time, measurements, five qualitative levels, improvement and prediction)

#4-Super-natural (objective vs. subjective, affirmation vs. negation, lawful vs. relative; invented vs. revealed, and various hybrid positions).

More information on this can be found in the *ECID*.

As described earlier, conditions existing during the 20th and 21st centuries have reduced the meaning of terms like 'science' 'knowledge' and 'reality'. Coupled with our information-saturation and prevailing marketing techniques, society can easily find itself participating in a 'virtual reality'. Effective marketing strategies and legally enforced taboos have helped re-write history. Millennia of historical qualitative values that had propelled our civilization to great heights are challenged continuously. Questionable subjective standards that throughout history had been at the root of failed civilizations are today being legislated as relativist social concern and truth derived from the 'democratic reality'. All this is done in the name of 'science' and 'democracy.' A democracy exported as an ideal, and forced upon other nations, has become the bedrock for instilling principles of failed civilizations.

These developments touch upon the underlying nature and principles of what constitutes the 'scientific method.' Imperceptible redefinitions (mutations) introduced through 'reductionism' and through the loosening of historic anchors have given way to five areas that have been re-examined in this book:

1.　Knowledge-based infrastructures and 27 management plans that reflect 'conditional certainty.'
2.　Decision-making and problem-solving tools for managing 'uncertainty' until the solution can reflect measureable conditional certainty (#1)
3.　Five levels of qualitative infrastructural change management
4.　Mental models – mechanical, organic, processive, information engineering
5.　8-point dynamic decision-making and problem solving steps
6.　Qualitative and geometric tools

4.3 - Man's Brain's Prioritizing Capacity

Myth is a manageable resource. As such, one can conceive of an intellectual think tank where all myths are created, designed ritualized, tested, documented and marketed. Various groups of consultants, social activists and true believers launch these well-financed myths/denominations into the marketplace and monitor them for quality performance (customer satisfaction). Then after comparing the results with objectives, these are improved or discarded – as any other manufacturing process or product. After all, in a social and business management environment visions, myth, religion and ideology are equally manageable resources that function together with economics, products, human resources, technology, demographics, and politics, etc. Each is a resource.

The human brain media manages media myth. The human brain is a management and decision-making tool that daily seeks certainty in an environment of uncertainty. Once a level of certainty is attained, there emerges a sense of comfort, stability, satisfaction and even happiness. In a condition of uncertainty the human brain uses the decision-making and problems-solving process and other tools to help resolve paradoxes and implement solutions.

The brain must eventually prioritize information and possible alternatives. The brain will filter these into a hierarchy of priorities. Inevitably, having identified the highest priority, the mind will de-prioritize all other priorities and values. Inevitably, due the management task of this complexity, the mind in the process of managing the millions of decisions uses the dynamics of the 3D Management Model, or works with parts of this Management Model. Much of the content, priorities and management methods are simply accepted (education), researched, optimized or reduced. However, this highest value remains as a guiding light and has us defend it with our lives.

This highest priority and value is derived from experience and from the environment. This highest priority/value, leads one to establish 'sanity' and security borders. As an information management tool, the brain automatically, converts this highest priority into levels of action (see Figure 5):

1) The Executive level (which helps formulate the purpose, objectives and strategies)
2) The Supervisory level – establishing: a) policy – functions of the intellect/reason; b) procedures – functions of the will; and the c) data/information base, i.e., rules & regulations – functions of the emotions)
3) The Functional level (where it can be used to perform research and development; economics; and deliver productivity)

With these 3 management levels and 27 plans, it is now easier to understand the content of the terms:

1) Religion, ideology and faith (S1-9) (highest priority – purpose, objectives and strategies)
2) Ritual and practices (O1-9)- operations management, policies, procedures and rules
3) Ethics (S1) management style, practices - management functions (D1-9), research and development, economics and productivity.

This 3DMM provides us with 'awareness,' a 'frame of mind,' 'identity' 'authority;' it also helps us identify ourselves with roles and scripts (self-image) and an 'attitude,' towards almost anything, and reflect a 'culture,' and all other Management Style plans.

All highest values/priorities acquire superlative, overriding, executive or '*divine*' qualities. This is why all ideologies and religions are fundamentally theistic - management of the highest priority/value. The skeptic, atheist, uniformitarian, myth promoter, believer in a myriad of gods, animism, economics, materialism, etc. all reflect a highest priority. This highest priority acquires an overriding 'divine' character or quality.

We underestimate the ancients who have identified their highest priority as 'gods.' These galleries of highest executive priorities imply a complete supporting management system. Epicureans or Stoics have their respective highest priorities (small letter 'gods') and a philosophical/management system that helps sustain their highest priorities. From this emerges a hierarchy of values (reality, ethics) and decision-making, problem-solving means for interpreting and solving challenges.

Even the Theory of Evolution reflects a pantheon of five gods (highest values): Dionysus (chaos) + Fortuna (chance) + Chronos (time) + Nike (victory – i.e., survival of the fittest). The ancients perceived that these four gods (Dionysus) compete with Apollo - form & art – the creative features – that uniformitarians express as statistical progression in isolated and challenging environments.

The ancients recognized that each and collectively, these gods reflect their corresponding management plans and levels (see 3DMM). Once we recognize our highest value (god), then our 3DMM gets into gear to accommodate us to our environment.

Similarly, it should be remembered that in spite of the spin that the media, academia, social science offers about the 'scientific certainty' of the uniformitarian simple-to-complex, in the past the same evolutionary views among the early Greeks had pure mythological interpretation. Greek philosophers[lxiii] - specifically Aristotle and today's pantheistic religious worldviews[lxiv] have not changed. We see this in such expressions such as - evolution 'plans', 'changes', and 'decides'. This, of course, may be a simple metaphorical expression, but we also find that at the United Nations, there are plans to ensure that all people would 'believe in evolution.'[lxv] This conviction that evolution 'lives,' and through consensus we find that most modern scientists 'believe' that evolution is certainty, this has came to influence even mainstream Christian Churches and denominations. The mighty Roman Catholic Church and many Christian Orthodox Churches have succumbed to Theistic Evolution. Pius XII (*Humani Generis*, 1950) identified a split within the Roman Catholic Church. So much so, that Pope John Paul II, had been influenced by and had been immersed in the writings of the Modernists, has succumbed to the Hybrid Uniformitarian theories (see *ECID* for more detail).

Where and how did modernist uniformitarian views originate and progress to the modern world? There are historical events, church authority practices, demographic changes that have brought about the influences. Materialist positions that led to the uniformitarian thesis posed challenges in the 14-16th centuries in West Europe within a context of a hidden agenda. It is a 'hidden agenda' because, if, for example, some quasi-protestants did not like some Papal administrative practices, the question is why hadn't these Protestants become Orthodox Christians – a religion that existed in Western Europe among the Celts.[lxvi] This alone is a separate question, but it is generic to the 'secret agenda' of the times (see *ECID* for details).

4.4 - Algebraic vs. Geometric Constructions of Reality

During the past 170 years, there has been an enormous amount of history rewriting. Civilizations, concepts and meaning of words, concepts, syntax, and stylistics were stabilized and defined within the principles of geometric natural law and within the

Christian knowledgebase. This approach provided workable, refined and measurable definitions that were used in scientific development, economics and management. In the absence of these laws, language predictably tended to become relative and subjective.

This suggests that communication and civilization reflect two possible basic mathematical constructs. We can trace two mathematical constructs that are at the heart of all philosophies, ideologies, religions, and beliefs. There are the algebraic and the geometric fundamentals that underlie all words, grammar, syntax, stylistics, metaphors, symbols, concepts, etc.[lxvii]

The algebraic world[lxviii] (Figure 19 and 23 below) begins with a 'point' from which the mind constructs/calculates all-rational possibilities. These can be expressed as symbolic extensions, for example – $a^2 * b^2 = c^2$. The algebraist's mind fills the 'points' through the required extensions and appear as a subjective and relative reality (i.e., unbound by external law). This reality may be pluralistic, democratic or collectivist that reflects algebraic reality. Each of the algebraic variables 'a' 'b' 'c'… may express theistic, non-theistic or materialistic realities. The pantheistic worldviews, therefore, emerge from the algebraic approach.

Figure 19 - Algebraic Reality

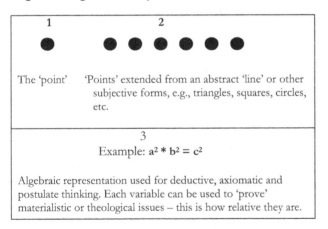

The alternate mathematical view that stands in contrast to the Algebraic worldview, is the Geometric approach (Figure 21 below). The geometric view does not begins with a 'point' but with 'infinity' - geometrically represented as a 'circle' where there is no beginning, end, numbers, nor extension. When the circle is 'folded' we see that circle forms a diagonal AB. This AB represents a sub-eternal condition from which come time, space on either side of the diagonal AB, and quantity - the two areas on either side of diagonal AB (time). By proceeding further – i.e., fold the circle perpendicularly to line AB, a second diagonal CD emerges. Where two diagonals (AB and CD) cross, we have 'point' E.[lxix] From this, creating a triangle is easy – fold side B to the point E forming lines FG. From F and from G make two separate folds connected with A. thus forming the equilateral triangle AFG.

Figure 20 - Geometric Reality

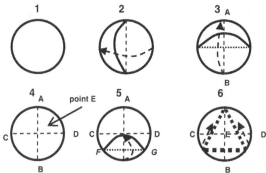

This brief construction exercise shows that the geometric eternal circle is the foundation from which one defines all of reality. The history of everything that folds reflects the 'inheritance principle.' In other words, each geometric configuration within the circle is a base upon the construction of a previous one. So you can trace one geometric structure through a series of earlier formed foundational structures - triangles, pentagons, octagons; the geometric solids (see Figure 21), as well as, the three-dimensional cone (see Figure 22) with its qualitative infrastructures, which evidence higher energy- densities.

The algebraic 'point' (Figure 19) is not independent from the infinite circle (see Figure 20, #4 'point' 'E'). This algebraic 'point' is contingent on five conditions within the geometric circle. It becomes clear that any human algebraic construct that denies the existence of these five pre-conditions, automatically become subjective and relative constructs. Any decision-making 'highest priority' or value that is conceived by a human *temporal* mind, while excluding the circle's (infinity) pre-conditions, automatically establishes algebraic sub-infinite – i.e., temporal conditions. Thus, priorities derived from this algebraic and temporal condition are at best pantheistic. Pantheism is the deification of nature / temporal existence.

The human mind's highest priority acquires deified qualities. This occurs because this highest priority derives executive management purposes and legal functions. Whether the highest priority/value is theistic, polytheistic, animalistic, ideological, and materialist - all of these algebraic relativist views reflect essentially the worship of a sub-infinite – i.e., a temporal (Figure 20.2) priority rather than infinite priority (Figure 20.1). Thus, all temporal highest priorities are essentially pantheistic. This also means that all other algebraic (non-geometric) pluralistic and relative highest priorities – whether they are theistic, polytheistic, non-theistic, animistic, philosophical or mystical worldviews essentially derive from the pantheistic worldview. This even places many of the denominational and non-denominational Christianities into this Pantheon of pluralistic

Figure 21 - The Geometric Solids

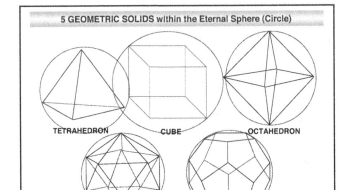

cults, because these Christianities adhere to the algebraic rather than the geometric foundations. Similarly, there are those who claim that they 'recognize' the existence of infinity, yet provide an Algebraic definition. Certainly, it is possible to talk about 'general' infinity and use Euclidian methods to talk about parallel lines extending into infinity. This in reality simply becomes an abstraction, a 'creative' leap of the mind or faith. The real geometric infinity must be described with the recognition of the inheritance principle, and with the proper positioning of the self and one's environment, 3DMM, DMPS, 8-Point DMPS / Scientific Method / Learning, because the Geometric Natural Law-thinking process is the 5th Model of Thinking.

The Christian Bible uniquely among all world literatures reflects this lawful geometric system (see Fig. 22, 23, 25, and 27).

As mentioned earlier and worth repeating again, is that one of the reasons why original Christianity could not have been made to fit into the 'pagan Roman Pantheon of Cults', is due to Christianity's geometric natural law. It identifies infinity of the true Supra-God (Supra-Monotheism) that supersedes all other temporal algebraically derived pantheistic gods (whether they are theistic or secular). In other words, materialism, atheism and agnosticism are nothing but reflections of algebraic highest priorities – derived by the temporal mind of man – from which we derive the original pantheism. Let me clarify this in the next two paragraphs.

It, therefore, becomes necessary to determine the character, 'nature' and framework of the algebraic and geometric priorities. Which one of these mathematical methods – algebraic or geometric, really helps us define the scientific method? What are the pre-requisites, facts and evidence presented by these mathematical models? How does man's decision-making and problem-solving mind interprets reality that these mathematical models suggest?

4.5 - Algebraic or Geometric Scientific Method?

Perhaps the best way to compare the Algebraic and Geometric approach to the scientific method is by examining basic Figures (16, 22, 23 and 24) - their explanations that are at the heart of the scientific method definition.

 Figure 22 and 24 describe requirements for establishing or identifying qualitative (vertical) change. These provide examples of infrastructural scientific and technical requirements for qualitative change. With these, we can forecast events and processes at various infrastructural levels of energy density, temperature and power magnitude levels – in areas of energy production and control (from wood to fusion). These same Figures are evident in other multiple ways and uses – e.g., in identifying weather patterns on Earth, planets and movements on stars; emergence of new structural development under high level of rotary velocities, software programming that involves robotonics. Inevitably this will include workings of executive faculties; not only quantitative but also qualitative economics, and as it will be shown below – in qualitative development of scientific knowledge and research.

The five energy-density levels can reflect realities in many fields – see Figure 22 and its corresponding Figure 17:

Figure 22 - Qualitative Infrastructures & Energy Densities

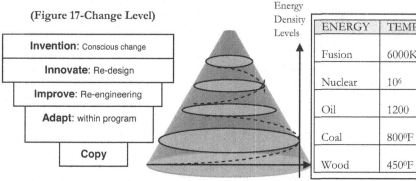

ENERGY	TEMP.	POWER
Fusion	6000K	Magnitude 2,700
Nuclear	10^6	10^8K
Oil	1200	4
Coal	800°F	2.4
Wood	450°F	1

The first energy-density level is copy. Students, scientists and leaders simply duplicate what they have learned (decision-making and problem solving, procedures, rules & regulations, processes, behavioral patterns, etc.). They become 'A' students, award winners, certified professionals. The aim is to maintain and apply existing knowledge, standards and thinking. Students learn, are tested and demonstrate that they can 'copy.' Scientists follow procedures and leaders maintain a steady course.

The second energy-density level identifies gaps – the difference between 'what is' and 'what ought to be'. Such comparatives search for adaptations and re-alignments. Teachers teach students, scientists improvise and leaders refine processes in using existing resources to help overcome anomalies, barriers that appear in existing systems. They seek to meet new quality, performance and productivity benchmarks. Students, scientists and leaders with unique skills, re-examine the knowledgebase for opportunities. Instead of 'reducing,' 'relativizing' or jumping into the fantastic blue while redefining existing localized processes into a new language, the true re-aligners use the knowledgebase to investigate, compare, document, create models, predict and test results to discover new laws and infrastructures upon which a new technology may be erected.

The third energy-density level involves improving, re-engineering. Some, while performing second level adaptations and re-alignments, also set up the groundwork for improving or re-engineering resources, systems and strategies. A global, supervisory and hypothesis (alternatives) based focus leads to qualitatively superior, faster, higher capacity mechanism and systems. These are strategic questions that help re-engineer resources, performance, more accurate forecasts and advanced planning and scheduling. Proactive students, scientists and leaders anticipate novel measureable qualitative approaches and discover new laws and infrastructures upon which they and their followers erect a new technology.

The fourth energy-density level includes innovating and re-designing. In a fast-paced environment where re-engineering is a standard, the student, scientist and leader soon realize that re-engineering is but a sub-component of re-design. Multiple trends reveal new patterns of packaging flows, information technology. For example, sub-atomic components, flight and submarine navigation have similar and unique process and infrastructural features. New engine, information and automation designs help achieve significant variations on economy of scale – more bang per buck. Supervisory and executive systems help address and further lead to new laws and infrastructures upon which they and their followers erect a new technology.

The fifth energy-density level is - invention or consciousness. Within this context, re-design becomes a sub-component of invention. This is where all aspects of the knowledgebase, all geometric natural laws are re-examined for executive level refinement. Students, scientists and leaders decipher the refinements within geometric natural law.

There are counterfeit conditions for each of these levels. An improver may consider him/her/itself to be an innovator and may cause havoc at inappropriate infrastructures

- while targeting the knowledgebase, the DMPS and the legitimacy of an authority. Social and political workers reduce human beings to 'image of primates' and apply herd and genocide economics (see 20th century events). While using 'improvement' terminology, the improver may wonder why no one understands his claimed 'innovative' concepts. By using second level causalities, misinterpreted evidence, comparisons, historical parallelisms the subject matter never moves out of 'uncertainty.' Organically minded scientists, administrators and scientificists begin to redefine 'science,' within organic models and ignore 'light-year' information complexity, leading to ideological authoritarian cultism. In such cases, such 'quasi innovators' ignore history and the valuable knowledgebase information and knowledge, re-write, re-procedurize, re-standardize interpretation and definitions.

Although such conditions can be found throughout history around the world, one must be careful with the sources and 'authorities' that interpret these events. For example, the Protestant Reformation, Freemasonic and Talmudic sources have frequently targeted the Roman Catholic Church (RCC) conduct and strayed far in their interpretation of historical and knowledgebase accuracy. Casting the RCC in an arch enemy role, the three justified their right to introduce 'adaptations', 're-alignment,' 'improvements,' and 'innovations,' while misinterpreting or ignoring the existing historical knowledgebase. Until that time, this knowledgebase represented a workable conditional certainty. Besides the perception of flaws in the theological position, there was a recurrent reductionist theme of historical events. For example: 1) the case of Galileo vs. Vatican; 2) purposes for the Crusades; 3) Spanish Inquisition, which allegedly exterminated thousands to millions of victims. I discuss these three issues in detail in *'Evolution, Creation, And Intelligent Design: Which ones are scientific?'* (*ECID*) (2009)

Such an approach to the last 500 year-re-written history, and strategy to disqualify the knowledgebase had also been staple source of information for decision-making in Soviet Russia. Here we have identical preconceived uniformitarian notions of 'primitive vs. modern knowledge, cultures, societies and civilizations (e.g., Victorian morals). Here, just as in the USA and Europe, if someone were to question the modern uniformitarian establishment and modernism, such questioning would be attributed to a wish to promote the primitive and pseudo-scientific world.

Uniformitarian interpretations of history, morals, psychology today permeate all areas of civilization. The uniformitarian notions of 'primitive illiterate-to-modern man' help socially re-engineer the democratic masses. Through subtlety, the masses are embedded with the compulsion to comply with uniformitarian standards, requirements, roles and scripts. Uniformitarians drill the set pattern of 'Evolutionary/Uniformitarian Realism' (ER and UR) 24/7. This is evident beginning with children's shows, K-12-college education, the media that work together to adjust our thinking. The *modern and tolerant* 'good' and 'free' is contrasted to the primitive, unscientific, intolerant and warlike past. We now think in UR/ER terms.

These examples show how and to what extent the five change levels in education, scientific community and leadership have reversed the qualitative 3DMM infrastructures, gap depth determinations – soon to be extinct.

By presenting these Figures, Tables and options, it will be possible to determine the distinction between what constitutes science and ideology, identify incremental increase in an accurate knowledgebase, and how to resolve uncertainty through DMPS and lead to conditional certainty.

4.6 - Algebraic Subjective Realities

Figure 23 provides the pattern or formula for establishing paradox and futility. The algebraic mathematical method cannot logically account for at least 24 conditions that must be linked through lawful relationships. This method, however, can provide a structurally fragmented, subjectively 'assumed' relationship.

Algebraic variables may be used to represent materialist as well a mystical concepts. These and every condition in between represent subjective and relative conditions. They are subjective because there are no objective foundation or reference points to anchor upon. Variables a^2, b^2 and c^2 can represent anything that the conceiver wishes the variables to mean. If the conceiver musters enough authoritarian power, then the conceiver becomes an 'objective' enforcer, until he/she/it is replaced by another relative enforcer with subjectively derived definitions and variables. These variables are not anchored on objective, lawful, verifiable, testable systems. A conceiver may state that variables in materialistic variables (e.g., uniformitarianism) that can be described, measured, tabulated, and quantified, to become 'facts'. But these 'facts' reflect a framework that will be subjective and tentative – i.e., hypothetical or theoretical – thus remain in a realm of 'uncertainty.'

This represents a fragmented approach that moves with the waves and whims of authority. Truth will vary with authority's whims. Today something is true, but tomorrow you had better adjust yourself to reject it as error because authority has now established a new truth. An algebraic approach is reductionist – seeks to provide lowest common denominator - mechanistic explanations for complexity, and is fragmenting because under such circumstances any mechanical explanation should do to explain complexity. Users of the algebraic method reduce and fragment reality. The reductionist will exhibit symptoms of amnesia and perpetually live in a world of relative and uncertain reality.

The algebraic approach, limits the content of the 3DMM (Figure 5) by functioning from the supervisory level (policy). It also eliminates the stylistic infrastructure and the executive management plans. As established above uniformitarian naïve realists 'UR' view a synthetic history through the 4-point uniformitarian notion – a) present is the key to understanding the past; b) simple-to-complex, c) 'primitive-to-modern', 'cultist-to-enlightenment,' and 4) closed materialist system. URs and ERs filter the historical knowledge base through these 4-point glasses only. It was, therefore, easy for the Marxists in the USSR to develop their doctrine of 'Socialist Realism' because uniformitarian and reductionist materialism were pre-established. Similarly in the West,

Figure 23 - Algebraic Subjective Rules

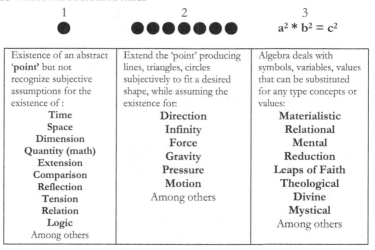

1 ●	2 ●●●●●●	3 $a^2 * b^2 = c^2$
Existence of an abstract '**point**' but not recognize subjective assumptions for the existence of :	Extend the 'point' producing lines, triangles, circles subjectively to fit a desired shape, while assuming the existence for:	Algebra deals with symbols, variables, values that can be substituted for any type concepts or values:
Time **Space** **Dimension** **Quantity (math)** **Extension** **Comparison** **Reflection** **Tension** **Relation** **Logic** Among others	**Direction** **Infinity** **Force** **Gravity** **Pressure** **Motion** Among others	**Materialistic** **Relational** **Mental** **Reduction** **Leaps of Faith** **Theological** **Divine** **Mystical** Among others

the overly optimistic 'vertical leaps' through simple chance, mutations, survival of the fittest in a challenging environment – has also been applied to all social programs, psychology, education, interpretation of law, scientific method, etc. The doctrine or policy of 'Socialist Realism' affected not only how reality was to be portrayed in the arts (literature, visual, plastic, music), but, as presented earlier, also in all of education, media, and daily conversations. A Socialist Realist must '*identify in today's world those elements and actions that represent the best examples of what will exist in tomorrow's communist world. Also, degrade all that will not be in the future.*' Similarly, this rule is valid in comparing the re-defined primitive elements of the past (e.g., Christendom) with the re-defined advanced elements in the present.[lxx]

Since algebraists are relativists, subjectivists and do not recognize universal, scientific and geometric natural laws, but only 'authorized' rules (e.g., NAS's), procedures and policy, then the scientific method and DMPS will be subject to the same reductionist applications. For example, the 8-point DMPS / Scientific Method will be filtered through the same 'Evolutionist Realism' criteria – producing and leading to a formula of failed civilizations.

Instead of adhering to 'engineering' laws that pursue knowledge by resolving uncertainty through DMPS tools, the content of Failed Civilizations extend and guarantee the perpetuation of 'uncertainty.'

One can infer from the above content that if people conform to the American Constitution and do not conform to the uniformitarian worldview, then judges, politicians, educators, corporate executives will see such as producing rebellion, counter-revolution, cultism, and racist supremacism. This is so because such an approach is derived from the notion and belief that man is created in God's image (supremacy aver primates?). UR, ER, SRs are categorically established upon the premise that human beings have evolved from primates (lower class).

Table 9 - Formula for Failed and Successful Civilizations

Reference: Figures 11, 13; Tables 4, 6, 7, 8 and others (see below)

SUCCESSFUL CIVILIZATION	FAILED CIVILIZATION
1. User/civilization refines all aspects of DMPS. Recognizes the '12 limitations of Science;' no limitations to alternatives and processes the hypothesis and theory through DMPS phases. A Successful Civilization recognizes all aspects of the 3DMM – 27 plans, 3 x 3 rows – executive, supervisory, functional, at the three infrastructural levels (design, operations and style). This includes the directive, processive & info-base columns.	1. Guided by the 'Luxury Syndrome'-filter pre-determines DMPS process – psychology, definitions, alternatives; from the filtering phase. Excludes 3DMM's historical value; reduces the 3DMM's executive plans and Stylistic level plans - infrastructure. Pre-determines theory and hypothesis and maximizes the scope of 'uncertainty' within DMPS. Reference: Tables 6 and 7; Figures 9, 10, 12, 13
2. For example, the Creation Scientific Model accounts for multiple alternatives. This also includes legal historical documented evidence that includes interpretations and predictions within the scope of a recent history, and up to 3 singularities	2. For example, the Uniformitarian Scientific Model arbitrarily filters reality and history through concepts that the present is key to interpreting conditions, processes, and rates that existed in the past; statistical progression - simple-to-complex and closed systems. Reference: Table 4
3. Maintains an Open System derived from Geometric Natural Law; also in compliance with OUSM.	3. Maintains a Materialistic closed system. Excludes OUSM and promotes UR, ER, and SR - naïve realism. Reference: Figures 19, 23
4. Maintains Geometric natural law; inheritance law. Authority derived from these that provide support to ensure that uncertainty converted to conditional certainty.	4. Maintains an Algebraic reality. Axioms theorems, a-priori; Authority based upon these rules – e.g., NAS in areas of sciences, and a-/Gnosticism in philosophical issues. Reference: Figures: 16, 20, 21, 22, 23, 24, 26
5. Enhances search and provision of means for infrastructural conversion methods	5. Will be unable to identify or distinguish between linear and qualitative change methods at five qualitative levels of change. Reference: Table 3
6. Seeks to identify, design, maintenance, calibration and	6. Will maintain a reduced or truncated view of management plans, rows, infrastructures and

management of all 3DMM processes	columnar directive function. It will also truncate human mind modeling; remove executive plans and stylistic infrastructure from the 3DMM. Prioritize supervisory functions (e.g., behaviorism). Reference: Figures 19, 23
7. Will use processes and gap depth determinations as tools to help convert uncertainty to conditional certainty.	**7.** Will oversimplify the 'input-processing-output-feedback' system, and limitation of the Gap Depth Determination, 18

What primate origins suggest is that mankind doesn't have an Executive management infrastructure (see 3DMM – Fig. 5) for establishing the Purpose (law, identity, scope) (D1); Objectives (authority, priorities, standards) (D2); Strategies (resources, forecasting, performance and planning) (D3), nor any of the Management Stylistic plans. In other words, human beings must recognize that they are: a) nothing but beasts of burden; b) proletariat vs. bourgeoisie/capitalists (Marxism); c) inferior vs. superior races (Nazism). This is a slap in the face of Christendom. This also suggests that only uniformitarians can rule these 'developing' masses. Following WWI and WWII allowed UR/ER to take over all Christian lands and governments not because uniformitarians and evolutionists have proved a true science and scientific method, but by introducing failed civilization values.

This is why the uniformitarian (scientificists) high priests wage war against the original definition of Christianity and the American Constitution. These Christian and American Constitutional executive-based belief systems suggest that man is endowed with inalienable rights given by Supra-Monotheistic God. The Christian Bible specifies that there are two kingdoms – one for citizens in the Kingdom of God and the other for those in the Kingdom of Babylon (Dark Age).

4.7 - Geometric Objective Lawfulness

In contrast to the algebraic pantheistic relativist method, the geometric method suggests not only the true foundations for the scientific method through all of its DMPS, 3DMM and other methods, but also expands the realm of inquiry based on lawful principles (geometric natural law, principle of inheritance).

One can conceive that there may be two approaches to the geometric natural law:

1) Derived supra-monotheism (highest priority, value) from the geometric approach[lxxi]; and/or the

2) Revealed supra-monotheism as uniquely depicted in the original Christian Bible, which reflects and is described in terms of geometric natural law from the Books of Genesis to the Revelation. The original Christian Bible is the only historical document

that describes the geometric natural law-based supra-monotheism. Other theisms and atheisms reflect algebraic and supervisory 'causalities.'

Both the derived and revealed approaches demonstrate the inherent scientific approach and method. The Christian Bible for example, stands in significant contrast to all other literatures and records - such as the Babylonian Talmud, Greek philosophies, Sanskrit, and other world literatures.

The Christian Bible contains unique features: it is –

A legal document with an Old and New Testament that require a testator and inheritor

Two legal covenants of marriage between the prime legal party (the Eternal Lord God/Jesus Christ) and the second legal party - genealogically tracked through the parties - the righteous, who adhere to the covenant terms, conditions and prophecies.

The Prime Party makes additional covenants and/or agreements with: 1) prime individuals and families that lead to the marriage covenant of the genealogical line; with some who are in the proximity and are not of the prime genealogical line (e.g., Esau/Edom, Ishmael, Nineveh - Jonah, the King of Babylon)

This documented knowledgebase (D) identifies a 3DMM design management structure. From this, one can derive the operational and stylistic management plans, links and infrastructures. This 3DMM represents a typology from which other realities can be constructed: psychology, national identities, the identity and mission of the Messiah, history, and others – see 'Tabernacle in the Wilderness'[lxxii] and partially in the 'Temple of Solomon'[lxxiii].

The design helps the righteous to:

1) Manage the Kingdom of God, which is contrasted with the anti-type Kingdom of Babylon contract (algebraic pantheism)

2) Determine psychology (a) executive functions – spirit; (b) supervisory functions - psyche (soul); and (c) functional (physical – search, development, economic, productivity)

3) Identify the responsibilities, functions of executives, priests, authority, supervisory, functional, etc.

4) Provide a document that records a 4000 to 6000-year history, precedents, case histories, civilizations, forecasts, economies, applications based on geometric natural law (e.g., Decalogue),[lxxiv] identification of the righteous personalities and multitudes of people who carry the original Kingdom Covenant and Christian message to the end of time[lxxv]. The Christian Bible's stylistic uniqueness and originality has not been

duplicated or surpassed in any world philosophy or literatures. This includes a combined: narrative style, poetry, total musicality from Genesis to Revelation, mathematical stylistic patterns (this does not refer to the Bible Codes).

5) The first chapter of the Christian Bible reveals a legal text. Beginning from the first verse and continuing until chapter 2:3 reflects geometric natural law as the self-identified Creator creates a canvas with underlying fibers. The Creator creates His highest creature - made in His image.[lxxvi] This 'image' is the executive functions and creative capabilities. Man identifies the Creator (prime party to the Covenant) as the 'Eternal Lord God' in Genesis 2:4 – identifying Him by his eternal nature - His supreme governing responsibility, and His nature, divinity, God.

6) The original letter-number value provides an additional layer of communication, music, relationships, identification of the various individual prototype positive or negative characteristics that are consistent throughout the Christian Bible.

All this and more, highlights the revealed portion of the Christian Bible, which may further show the relationship on how: 1) the first six geometric process figures match lawfully with: 2) the corresponding first three verses in the book of Genesis – see below and also Figure 26. This suggests lawful elements in the scientific method and scientific evidence.

4.8 - Comparison of Geometric Law, both Derived and Revealed

Geometry derives from the Greek concept of 'land measuring.' It also extends to encompass all lawful referential relationships. In contrast to Euclidean/ Aristotelian (algebraic) geometry, established upon axioms and postulates that are derived through deductive methods of reasons; the Greek Platonic geometry ('non-Euclidean' geometry) relies on proof acquired through rigorous methods of construction. Originating within the eternal circle and tracked through the inheritance principle.

To better track events listed below in Figure 24, the reader should use Figure 20 – 'Geometric Reality' as a graphic reference to the process described below. In addition, Figure 24 lays out the initial steps - the first day of Creation that is described in Figure 26, which also continues to provide installed geometric solids within the remaining days of creation. These solids provide the infrastructures for what the Creator God creates each day.

Figure 24 - Comparison: Geometric Natural Law: Derived and Revealed

#	GEOMETRIC NATURAL LAW	CHRISTIAN BIBLICAL GEOMETRIC NATURAL LAW
1	The infinite dynamic circle, which has no beginning nor	There two translation versions for Genesis 1:1: a) 'In the beginning [time] God [infinite circle]

	end, numbers nor space, is the best rational construction that the temporal human mind can understand about infinity and from which time (a subsystem of eternity) is structurally initiated as a base for all other structures within eternity.	created...' b) 'God [infinite circle #1] in the beginning [time #2] created...' The second version (b) seems to conform to the geometric sequence (see Figure 20, #1, and then #2). The Biblical God provides His own name (God, the Eternal), nature and character (the Creator). No man observed this creation except God. In Gen 2 Adam sees God as the 'Eternal Lord God' describing the eternity, God's relational position to Adam's view – LORD God. Later, John (1:1) identifies another aspect of the Creator's name - the 'Word' – the formulator – also a management and organizing function. Later Adam will 'name' Eden's animals. This 'Word' formulates/creates within the infinite circle - out of 'nothing' i.e., out of infinity. Two conditions exist – infinity and time, and two of three personalities – the Infinite Father and the Word. When speaking with Moses, God identifies Himself as the 'I Am' (eternity) – the condition of permanently being.
2	Law is viewed as infinite geometric constructs - the infinite circle or sphere. Time emerges when the circle is 'folded,' with two areas formed by the temporal diagonal. From the managerial point (3DMM) of view, the executive plan reflects the purpose (D1) which is composed of Law, Identity and scope. Note: distinction must be made between Executive 'Law' (D1) and Supervisory 'rules' (D6) which are usually miss-defined as 'laws.' supervisory level plans. More specifically rules pertain to the database of information, while laws to the executive fabric of Creation. Note: because algebraists do not begin with the infinite circle, they commonly misapply the term 'law' ='rule' -see the difference in Figure 5.	Law is viewed in terms of geometric constructs - within the infinite circle or sphere (geometry), the diagonal fold (A-B) – representing 'time' now, by its presence, opens two areas on either side of the diagonal (time). This shows that as soon as time is created, quantity appears also – calculation of time relative to two areas – which is represented by the dual grammatical form. Here, geometric natural law is seen as an integral part of the very foundations of Creation. These created features – form the very fabric of all of Creation – God, the Word, infinite circle 'geometry', time, space, quantity - fibered within geometric natural law. Management kicks in as soon as there is more than 1 (3DMM). At the Executive level, (D1 thru D3) Law is an integral part of Purpose (D1). Here, within Purpose, Law couples with Identity and Scope. It is important to distinguish Executive purposive Law (D1) from Supervisory rules (D6) which miss-defined as 'laws.'
3	Divine Law can be understood by constructing, theorizing and hypothesizing, and by reflecting upon Geometric Natural Law (see below). Reflecting and extending the attributes of natural law can help approximate divine law. For	Divine Law – is revealed to man through the intermediary of geometric natural law and covenant law (Holy/ Sacred Scriptures & Traditions). Mankind, having been made is subject to geometric natural law, understands communication that complies with the Divine and natural laws. Man, with his temporal faculties cannot recognize the nature of infinity

	example, the nature of infinity (without beginning or end) must represent an eternal presence – i.e., the act of being, total perfection, wisdom, and knowledge since all is created and originates from, and no other source other than infinity. There is nothing in time that is un-knowable within infinity, because time is a sub-set of infinity.	and total Divine Law. Similarly, the features of divinely revealed law, His perfect creation and communication, are evident from Genesis 1:1 through Revelation 22. In fact, the content of Genesis' first chapter is written from the Creator's point of view since man had not been there from the 'beginning' (creation of time). Evidence of the 10 Commandments is clearly laid out from the first three chapters of the book of Genesis and concurred by evidence in other Biblical books. Note: there are algebraists who play with geometric circles, but the axiomatic algebraic circle is not set within an inheritance structure.
4	Natural Law – derived geometric law constructs become evident in, and become the very fabric in all of creation and the structure of and definition of reality and sanity. These are discoverable laws and are at the foundations of all mathematics, science, technology, engineering, government, society, history, psychology, communication and development.	Natural Law (law of nature) derives from geometric law constructs. The law of nature becomes evident in, and becomes the very fabric in all of creation and the structure of and definition of reality. These are discoverable laws and are at the foundation of all mathematics, science, technology, engineering, government, society, history, psychology, communication and development.
5	Natural Law – Language, communication, and mathematics are languages with their own: Words (data, information, concepts, and definitions); sentence (formulas); grammar (rules); syntax (specific formula constructs – sentences and rules: simple, complex, subordinates, compound, integral, etc.); Stylistics: a strategic arrangement to ensure clarity, accuracy and purpose of message).	Geometric natural law – reflects not only documented but also ascertained tradition. Some tradition is documented in terms of the original Christian Bible (scriptures). Documented transmission of Divine communication with man is evident from the earliest patriarchs down to Christ's apostles; through Covenants, agreements, testaments, history, management plans, genealogies, missions and forecasts. The Christian Bible tracks covenants and precedents, by writers in the genealogies of the righteous and historical events. As a legal document - it contains a starting point – the signing of the covenant by legal parties, case studies, judgments, statutes, history, strategies, policies, practices, rules, development, research, economics, productivity, quality standards, defense, science, engineering, sociology, psychology.
6	Geometric natural law and law of nature – the creation of time (diagonal AB - see #2 in Figure 20 – Geometric Reality) creates conditions of: 'Counting,' numerics, mathematics. Two areas of space on either side of the temporal fold diagonal (AB); extension; direction	
7	Geometric natural law – By folding the circle a second time to form a second diagonal CD (see #3, Figure 20), that is placed perpendicularly to the first diagonal AB. The geometric field lines allow for the creation of: comparison, reflection, texture, relations, logic	

| 8 | Natural Law – Identify Point 'E'. Where the two diagonals cross (AB; CD), we have 'point' E (see #4 in Figure 20 – Geometric Reality). This is a geometric force field focus within dimensional or spherical space. | Geometric natural law and law of nature –'…and the earth. [Gen 1: verse 2], now the earth was formless and empty, darkness was over the face of the deep…' (Gen 1:1-2). Some Christian authorities see in this passage the created original planet Earth, which was formless (whether: a) describing its surface or general mass itself, or b) a condition after some catastrophic event – that would help accommodate a long period of time). However, the geometric approach clearly shows that the formlessness and emptiness of 'earth' is the crossing (focus) of two geometric fields (folds) within dimensional or spherical space. An alternate translation for the term 'earth' in verse 1 can be 'firmness' – a condition that represents a 'firm' location (point 'E') where two folds cross (diagonal folds AB and CD) (see #4 on Figure 20). The second verse shows that the condition of 'firmness': Didn't have any shape (form) since it was a condition of two field foci. Was 'empty' ('firmness with no content of itself). Occupied a place in space 'where darkness was over the face of the deep/abyss (space)'. At this stage, this field/fold didn't affect space with any type of energy – (see verse 3 for this energy). |
| 9 | Geometric natural law and law of nature –Equilateral Triangle - Reference # 5 & # 6 in Figure 20 – Geometric Reality. Create an equilateral triangle, by folding and bringing side B to the point E thus forming line F and G (5). From F and from G lead two lines to A, thus forming the equilateral triangle AFG (#6); this suggests interplay of the existing geometric lines, to create a geometric 'force-field', that instrumentally form bounded gravitational forces and the preconditions for the electromagnetic spectrum. | Geometric natural law and law of nature — Equilateral Triangle - Reference # 5 & # 6 in Figure 20 – Geometric Reality. In Genesis 1:2 – '…and the Spirit of God (mighty wind) was hovering (moved) over the (the face of the) waters.' This is where the third person of the Trinity is introduced – the 'Spirit of God/the mighty wind'. This Spirit introduces: a) Movement – to move what was a static condition until now; b) To introduce change; c) Among the multitudes (waters) of 'points' – 'firmnesses' (earth) in the vast expanses of dark space. We see: d) Gravitational forces forming among the 'waters' (numberless 'firmnesses') coming together within the geometric lines and gravitational or field forces (geometric triangles) to create solids and bodies. It is with the formation of the geometric diagonals and specifically equilateral triangle (gravity and force fields) that the necessary pre-conditions appear for the creation of; e) Electromagnetism, sub-atomic, atomic and molecular hard bodies in space in Genesis 1:3 - the appearance of light.' |

This brief analysis of the first three Biblical verses, describes the beginning of geometric natural law. Geometric natural law continues and is evident not only until of the end of chapter, but continues into chapter 2. Both the 'derived' and 'revealed' explanations demonstrate the interweaving of geometric reality with its incremental development within geometric solids (Figure 21), and qualitative infrastructural and energy density (Figure 22). Beside these starting evidences, the geometric approach #1 through #9 (Figure 20 to 22) suggest at least 60 additional constructional derivatives:

10	Motivation and intent – wish to communicate and to listen		
11	Duration – continuous time (in the past, present and future)		
12	Tenses and aspects – simple 'exist' (-ed; will) includes the beginning and end of the activity – whether present, past, future); continuous' existing' (action continues as long as it is/was/will be done); perfect (have/had/will have existed' (indefinite continuing action towards completion), and other tenses.		
13	Beginning (start)	14	Process
15	End – (completion)	16	Quantity – numerical, relational, formula
17	Quality – infrastructural. & graded (Fig. 22)	18	Measurability – reference to standards
19	Dimension – size, volume, proportions	20	Comparativeness – now with n or x
21	Division – at least 2 areas on either side of line AB	22	Historical – past quantities and conditions – in inheritance principle
23	Forecasting – potential, future quantities and conditions		
24-40	Project management executrices: plan, schedule, contract, law, timelines, functions, activities, resources, controls, mechanisms, logistics, resource utilization, objectives, goals, policies, procedures, rules, development, economics, productivity, etc.		
41	Dimensions – at three +	42	Sub-and atomic matter
43	Molecular	44	Velocity speed – time, distance, rate
45	Pressure	46	Temperature
47	Forum-Structure-Design	48	Frequency
49	Calculus	50	Electromanetics

#51 - 70) +. In a separate study, the 7 days of Creation of the first week correspond to the entire set of Geometric Solids' (see Figure 26). For example plant life reflects the structure and dynamics of the pentagon-based icosahedrons, while man and woman (6th day), the seventh day of rest; Adam (eighth day of the second week) correspond to the highest life-based system reflecting a dodecahedron creation design. See: geometric solids (Figure 21) and the qualitative infrastructures of energy densities (Figure 22).

Geometric fields and solids (Figure 20) suggest that all reality is installed or composed of lawful geometric fields and infrastructural energy densities. They are built into reality's structures - 'fabric,' or 'canvas'. These geometric fields, solids and energy densities are evident not only at the sub-atomic, molecular, natural and animal life, but also within human life. Scientists are deciphering similar results in the solar system (see the works of J. Kepler), intergalactic, galactic clusters, supra-clusters, network of supra-clusters, new structure appearing within the high-speed rotary systems and at information processes and at the chromosome levels.

Clearly, the definition of the scientific method acquires substance in the integral geometric law of creation. It is not evident in the narrow subjective, algebraic and materialistic economy that cannot ascend above the empirical and a two-dimensional theoretical level, based on supervisory rules - uniformitarianism.

Returning to the journalist/editor/court's 1st question 'Do you think that teaching evolution goes against religious beliefs?' This one sided and ideological view is not a scientific question. First, one must address the implied 22 ideological inferences (see Appendix 1) before formulating a properly scientific question. Having addressed these many issues, one can properly formulate a more realistic question - 'Do you think that teaching pantheism goes against the original geometric law-based definition of the scientific method, and against supra-monotheism?' The answer is clearly in the affirmative.

CHAPTER 5 - COURTS, SCORE CARDS, FILTERS and SOLUTIONS

The second journalistic/ editor/courts ask is: 'State and Federal courts ruled in favor of Evolution's approach to science rather than that of Intelligent Design, Creation or Hybrid Uniformitarians. How do you interpret this ruling?'

5.1 – Uniformitarian Courts

Some US Federal Courts have interpreted that the Theory of Evolution is a religion-free approach to science. On the surface, this appears to be a safe interpretation and position to take in view of the alternative – a potential for a religiously and ideologically tainted science. Even though the court decision seemed to have addressed issues of science vs. religion (the demarcation line), as well as, of 'Church and State' issues (Establishment Clause), the decision did not address the 'Science-Ideology' relationship (Table 8), which seems to escape scrutiny in the courts, education and the media. Similarly, the Courts' provide reasons for their decision, yet it is clear that they ignore issues of a) 'objective and unfettered science and scientific method' (OUSM), b) geometric natural law (essence of the American Constitutional Law), and 3) 3DMM (scientific knowledgebase) and DMPS (decision-making and problem solving). Having promoted uniformitarianism instead of the founding American Constitutional law, too many Courts inevitably default and subscribe to the ideological notions that evolution is science and that Creation and Intelligent Design science are religion.

By posturing on this surface uniformitarian platform, Courts fail a second time, by not looking at what lurks beneath the evolutionary science platform. What is the infrastructure upon which the theory of evolution is erected? Evolution's infrastructure is really the 3 to 4-point uniformitarian ideology. At the same time, it is the four-century old pillars of: rationalism, empiricism, materialism, reductionism and modernism that support this ideology and this infrastructure. These five philosophical views oppose the principles and interpretation of the American Constitution, which has originally reflected geometric natural law.

The third flaw lies in that if these uniformitarian-based judges wanted to promote its ideology (its definitions, interpretations of what is science and religion) then these judges overlooked the opportunity to use the best uniformitarian model – atheistic Marxism. Methodologically, the Marxist uniformitarian approach is more systematic than the Secular Americanist one. Marxist uniformitarian is very atheistic (anti-religious), materialistic (closed system), reductionist (explains complexity in terms of mechanical, organic and processive models) and is economics-based. Marxist 'scientific' economics is just as 'scientific' as that promoted by uniformitarians (evolutionists) in America.

Yet it is evident that neither Evolutionists and/or Marxists have ever used the objective and unfettered science and scientific method, the 3DMM and DMPS. Instead, we see Marxist and Secular uniformitarianism perverting the scientific method by compromising it with UR, ER, and SR. If the scientific method (DMPS) is to convert uncertainty into conditional certainty, then it is clear that the uniformitarian approach ensures that all science remain at various phases of uncertainty.

The fourth court decision flaw, is that based on the above three deficiencies, the American Courts passed decisions upon something the judges couldn't define: a) what constitutes to be the original meaning of true science and the scientific method; b) what are the similarities and differences among ideology, philosophy and religion. This deficiency is evident because these judges have shelved American Constitutional law, which is derived from geometric natural law. This is a common trend among all the 'activist' judges.

The fifth flaw is that due to the above constraints, these judges did not help resolve through their decisions, the lingering questions whose mission this book addresses:

a)   Had the original scientific method been improperly defined?

b)   Had the method or the nature of science changed with time?

c)   Wasn't the scientific method supposed to provide a constant – an authoritative anchor?

d)   Has science and the scientific method been misapplied by some or all scientists?

e)   Has science's content been misunderstood, blended or mutated into economics, philosophy or ideology?

f)   Perhaps science and the scientific method had not been designed to fulfill a role that most thought it should fulfill

g)   Has the scientific method expanded beyond its procedural scope?

h)   Does science now provide evidence for scientific misconduct or intellectual dishonesty?

i)   Have influential forces hijacked science and lead it to fulfill other purposes?

The answer to each question is in the affirmative. Today, the historic perspective has been relativized. All standards reset to reflect 'naïve realism.' Uncertainty must reign supreme. Most areas of the American Judicial system are brought into this filtering process.

Two groups of scientists – the 'Young Earth Creationists' and the 'Intelligent Design' - had recognized the warning sign very early in this continuing debate. These scientists attempted to maintain historical filters. They updated their scientific models and pursued to maintain the original science and the scientific method. They went out to distinguish between uniformitarian and the objective-unfettered scientific methods and results.

Uniformitarians/evolutionists accuse that non-evolutionists 'find it difficult' or too 'complicated' to understand their 'evolutionist realism.' However, this books has proved that evolutionists seriously under-estimate the true conditions of competencies. The uniformitarian /evolution as well as their UR and ER represent 'naïve realism' and non-evolutionists understand the uniformitarian machination very well. Creation and ID scientists recognize the uniformitarian axioms, theorems and postulates. Creation and ID scientist clearly recognize that uniformitarian 'naïve realism' reflects mechanical, organic and processive mental models and fail at the information and the geometric natural law mental models. The 4-point uniformitarian principles: using today's events, processes, rates to measure the past; simple-to-complex or primitivism-to-modernism extrapolation is designed to sustain a materialistic worldview (ideology). We can see evolution, uniformitarianism, secularism in the concepts of: class struggle (Marxism), race (Nazism); division among genders, generations, promotion of genocidal programs such as abortion and Modernism (Americanism). Courts that base their court processes and decisions upon such uniformitarian foundations have legitimized, during the 20th century, social re-engineering plans that included concentration camps and genocide programs on a massive scale around the globe (e.g., Club of Rome). It becomes evident that objectives derived from the uniformitarian foundation are to convert the masses into controllable consumers and an economically manageable resource (proletariat, collectivism, consumer, etc.)

The disproportional battle that evolutionists wage is not so much to promote 'science' and the 'scientific method.' Instead, they target the Christian Knowledgebase, the developing application of true science, the scientific method, the 3DMM that includes executive and style infrastructures. What evolutionists aim is to 'reduce' the only means by which uncertainty (relativism, reductionism) can be converted into conditional certainty – in other words, evolutionists wish to introduce cultural blindness (3-4 point view) and amnesia (historical).

In contrast, the Christian 2000-year old creationist model has provided a practical legal and historical foundation that led to measureable solutions. It is Christendom that allowed civilization to revive and survive after several challenges: the Western Roman Empire began to falter (5th century), the 14th century's effects of the Black Death, and attempts of countless invasions. Christendom established a workable knowledgebase, standards, and several management systems to overcome irrationalism, reductionism, subjectivism and barbarism. It resolved these through measures such as the Augustinian concepts of the 'City of God,' (fifth century AD), the Byzantine Empire, Christian Celtic networks, the re-introduction of the Renaissance (15th Century) qualitative methods, and recognition of man's design in the 'image of God.' This opened the way to the development of engineering, sciences, technology, arts & letters, exploration and discovery. Here, the need to perfect man, society and civilization in the image of God's standards, lead to the rediscovery of geometric natural laws and the scientific method. This is what motivates the battle against the pagan theories and their program to re-introduce elements of failed civilizations and the dark ages.

The Intelligent Design group, emerged from the Neo-Darwinian group. With the assistance of high-tech research, the I.D. scientists discovered many anomalies in the theory of evolution's ability to predict and explain complexities at the DNA and Cosmological level. Without entering into the theological issues, the Intelligent Design group provided a scientific hypothesis of an Intelligent Designer that helped explain the new info-based complexities. It is ludicrous to see how uniformitarians address the complexity of information and nano-robotic engineering at sub chromosome levels. Most evolutionists still argue from the organic model – talking about gene duplication, mutations, and point to examples such as the nylon-eating bacteria.[lxxvii] Again, simply continue to present the trial-and-error, statistical progressive argument for information since this is a 'part the genome functioning within an environment, describing about the proportions of neutral, beneficial and harmful mutations'[lxxviii].

At the same time, I.D. scientists with their new approach had been able to continue with their scientific research without being stalled by artificial 'unscientific' uniformitarian realism (UR) filters and limited organic mental models.

Figure 25 - Comparisons: Filters - Evolution and Creation Science

UNIFORMITARIAN/ EVOLUTIONARY	CREATION SCIENCE
Gradual development within statistical progressive rules that exist today. Extrapolate existing physical rules and tolerances (e.g., speed of light) to an original explosive beginning (Big Bang) or a Steady State, and apply rules of simple-to-complex; primitive-modern. This provides a synthetic pre-/history to explain contemporary conditions	Conditional certainty recorded as a knowledgebase, which is legally documented as history (scriptures) – identifies geometric natural law, universal physical infrastructure, three singularities – first provides an 'open system' allowing for extra-materialistic participation or influence. Second singularity is a universal drop on an energy-density cone. Third singularity - an initial global hydro-tectonic

	catastrophe of short duration (12 months) and sequence self-adjusting global mega-changes over thousands of years leading to our current conditions, which are subject for empirical analysis, modeling and predictions
Materialism – economics-based reality	Geometric natural law – objective with inheritance tracking.
Reductionism – reduce all processes to mechanical, organic and processive levels	Qualitative energy-density infrastructures, re-engineering, re-design.
Survival of the fittest	Self-perfection through man's executive faculties and geometric natural law.
Genocide of any variety: classes (Marxism), race (Nazism), age, gender, ideology, abortion (Secularism)	Multiplication of population supported by applied qualitative technological invention and 'manufacturers' (Alexander Hamilton)

How do these filters function when it comes to establishing objective or 'unfettered' scientific method ('solving uncertainty') and knowledgebase (leading to quality assured or guaranteed 'contingent certainty?' Filters provide the scope for conducting 'interpretation.' The DMPS filters introduced in the fifth stage of the uncertainty process include three different stages of interpretation of objectively collected and recorded data and information. On the DMPS (Figure 13), these filters reflect: a) standards ('E') and b) history ('D') of the Knowledgebase. The filters reflect Management Design's Executive plans (purpose, objectives and strategies).

Where and why does the Uniformitarian method (Figure 14) deviate so badly? First, uniformitarians axiomatically associate philosophical and ideological concept of materialism with 'science'. In this process, science is re-defined not objectively but through a strictly *a-priori* materialistic perception. In this content, science is no longer the 'pursuit of accurate knowledge' but a guarded focus on the tangible, simple-to-complex progression, closed system that may be expressed in economical terms (Marxism). Here, the accuracy of knowledge plays a secondary role because the naïve realism first pre-determines specific parameters for psychological predisposition, definition, limited alternatives, set filter and conditional laboratory tests, confirmation and predicted outcomes. See the all-permeating UR/SR filters throughout the DMPS phases. In other words, materialism (and skepticism) cannot be understood outside uniformitarian principles. Any data or information that does not fulfill the prescriptive requirements will be considered to be: contaminated, inconclusive, irrelevant, or dead-ended. The same uniformitarian approach applies to social issues – this is why uniformitarian ideologies include mass genocide programs. Due to their naïve realism, they are unable to address complex issues.

This approach and reality is what inevitably the uniformitarian court judges promote.

5.2 - Report Cards

When evolutionists pontificate on what constitutes science vs. cultism, scientific method and reality, such uniformitarian approaches scatters scientists into several camps. There are the 'evolutionist realist true believers' - such as Richard Dawkins, who professes a 'methodological materialism.' Stephen Jay Gould remained neutral on issues of extreme materialism and theism. There are those who have pursued 'hybrid' positions; other subsets of 'theistic' or teleology positions (I.D.). Others play a progressive role by balancing long ages with 'creative reasons' (www.reasonstobelieve.org). There are those who may consider purpose, design, formal causation, progressive evolutionism. They allow for billions of years of pre-/history on the one side, and consider evidence for universal fine-tuning and physical constants of nature on the other (Denton). Some, among the latter group may also embrace universal common descent.

There are also the 'Young Earth Creationists' (YEC), who, while existing in this environment of re-definitionism and re-scientification, appear like an odd group that may be sitting on the fringes in comparison to others, but who have followers who number in more than half of the American adult population (60% - Gallup Polls)[lxxix]. Here, the Young Earth Creationists are the ones whose scientific model sustains, balances scientific evidence and helps predict systematically and successfully, and represent better the original millennial Christian position on Biblical Creation.

As examined earlier, the atheistic and hybrid uniformitarians see the Christian Bible through uniformitarian eyeglasses. They see the Christian Bible as they would any world literature. Such literature may have anthropological and ethnological value but in their uniformitarian view, this has little to do with 'science' (uniformitarianism). They see that such literature is strictly artistic, subjective and provides a mythological view of the reality. Yet, after removing uniformitarian glasses and look at the Christian Bible through the principles of Geometric Natural Law and covenant management, the visible picture becomes unique and highly scientific – including a 3DMM knowledgebase, DMPS and OUSM.

Again, why does Uniformitarian 'science' fail on the report card? Figures and Tables in this book identify items, issues and data for measuring the results. They contribute to the *Scientific Report Card*. In addition, this book offers textual content comparisons – the OUSM (objective unfettered scientific method (2.2), the '12 Limitations of Science' (Figure 14 and Chapter 2.6); DMPS (Chapter 2.4); 3DMM (Chapter 2), which includes 5 Thinking Models (2.7); Five Qualitative Change levels (2.8) and others that supplement these Figures and Tables. In all of these areas, at best, uniformitarians simply provide an uniformitarian ideological interpretation of science and reality.

In addition, this uniformitarian ideological interpretation is highly *deceptive* (as listed in the NASs 24 Rules). Uniformitarian English describes and reasons in non-traditional or cultural English definitions, meanings and management style (S1-S9). These uniformitarians use a language that is similar to that used by Marxist, Nazi, denominational and cultist users who truncate cultural, history and stylistic value and focus on one of possibly 1/12 lexical definitions and concepts that the historical language has.

The Christian Bible had not escaped reductionists' and uniformitarian's scrutiny. For over 200 years, proponents of 'Evolutionary Realism'(ER) have used the 'primitive-to-complex' concept and other uniformitarian modernist views to chip at Christendom's 3DMM historical database. Yet, despite this cynical chiseling in areas of linguistics, archeology, geology, dating methods, anthropology, etc., these and other circumstances provided 'last minute' discoveries and evidence that confirmed the Bible's described evidence and the erroneous assumptions and alleged theorizing evidence offered by the modernist (uniformitarians). These unexpected yet strategic discoveries are evident in linguistics, archeology (original Qumran documentation, translation of Egyptian, Assyrian, Hittite scripts), geology, dating methods (proving the authenticity of Old Testament renditions of events), anthropology (discovery of cities, towns, skeletons, pottery), CRAY computer simulations (see the above references – www.icr.org).

For example, some ER scholars suggest that there had been three sources of the Book of Genesis. Yet based on the legal, covenantal, testament style of the book of Genesis, we find that there are up to 10 pre-Flood and 14 post-Flood writers, patriarchs who left written records. Such authors have been documented: Adam, Seth (Gen. 5:1 to 5:7), Enosh (Gen/ 5:8 to 5:10), Canain (Gen. 5:11 to 5:13), Mahaleel (Gen. 5:14 to 5:16). The list of Who's Who continues until Prince Moses, who has Israelite (Abraham, Isaac and Jacob-Israel) extraction, summarized the voluminous knowledgebase, that had been carried around by the patriarchs from the time before the Flood, on a voyage on the Ark, and continued after the Flood until the time of Joseph in Egypt. This edited summary became the Book of Genesis written in Egyptian script. These writings reflected the Covenant requirements – genealogies, laws, statutes, commandments, judgments and precedents.

The first chapter of Genesis clearly reflects a text whose structure had been based on geometric natural law (Figures 24 and 26), resourced infrastructures and geometric solids (Figure 26). None of these key strategic evidences appear in the 'scholarly scientific' works, published by uniformitarians. Yet, the Prime Covenant maker, the Supreme God, writes Genesis 1 -2.3. This Supreme God and magistrate created all of Creation in 7 days (Gen 1:1-2:3). This first chapter contains a 'GANTT chart' of scheduled ('evening' and 'morning') functions, activities, tasks, resource loading, quality control ('this is good'), functions, activities, dependencies, contingencies, establishment of supervisory systems. At the same time, Genesis 2:4 through 4:26 identifies the legal writings of the Second Party to the Divine Covenant Adam, who identifies:

a) The Prime Contractor as 'the Self Existent' 'the Eternal Lord God.' Adam records what the Eternal Lord God granted to him - executive responsibility and authority over all of Creation

b) Terms and conditions of the Covenant, his CEO position and responsibilities within this Covenant

c) An <u>executive choice</u> (free will) providing two decision-making 'trees' – one leading to immortality and the second choice leading to <u>mortality</u> (spiritual creatures) and thus subjecting all that was placed under his executive authority to – energy starvation - death. Immortality reflects Jesus Christ's nature after the Resurrection that the second Testament and new Covenant describes. This is the purpose for the creation of man.

d) Having made the wrong executive decision, the now mortal (imperfect) CEOs had been expelled from their exclusive Executive Real-Estate - the Garden of Eden. Naïve realists, having rejected executive and management stylistic value, must reject such scopes.

e) Thereafter, Adam's son Seth (Gen. 5:1 – 5:7) continues to record the legal historical events, and Seth's Covenantal righteous descendants continued the record that had been transmitted to through Abraham, Isaac and Jacob to Israel's kings and prophets, appearing as the Septuagint version at the time of Jesus Christ's first advent. This contribution established and increased accurate knowledge – a legally managed 3DMM.

It is not surprising that after 200 years uniformitarian realist (UR, ER) scholarship, that had promoted 'Higher Criticism,' 'Modernists' and other superlative positions, uncovered this 'scholarly' initiative as being nothing more than that similarly promoted by the wizards of Babylon. They are blind and act like amnesiacs concerning the Christian 3DMM foundations and formal facts. Their reductionist method could not help them identify the Prime Contractor, the existence of a Divine Covenant, the divine and human abilities to write, record, decide and interpret issues that affected the condition of all of creation, the true descents of man - made in the image of the Eternal Lord God. Instead, these modernists theorize about modern man being made in the image of primates, through chance, through the struggle for existence, and within billions of years of synthetic historical fiction. These neo-interpreters prefer to promote myths derived from the 4-point uniformitarian model. All this clearly, disqualifies them from conducting 'scientific' interpretation of the Christian Bible. Their pseudo-science has permeated every education curricula and certification programs.

If the Christian Bible were to be examined within the OUSM, DMPS and 3DMM parameters then it is easy to identify geometric natural law-based scientific foundations. The 'Young Earth Creationists' (YEC) created a modern 'Creation Science Model' in the 1960's, to process all empirical data for proper interpretation and predictions. The YEC designed and set the Creation Science Model within the parameters of the recent historical events as recognized by the early Church Fathers and two millennia Christendom. This included the legal historical documentation of three recent singularities. They used the original understanding of science (pursuit of accurate knowledge) and have been refining the decision-making and problem-solving method (DMPS) over time. With this, the Young Earth Creationists and their Creation Scientific Model succeed on multiple items on the 'Scientific Report Card.' The YEC does not distort the fundamental framework of objective or 'unfettered' scientific method (OUSM).

This is evident when YEC:

1   Examine records and publish evidence empirically (geology, biological, fossil record, dating methods, the cosmos, etc.
2   Construct, describe, test and predict data through a scientific model that allows for a worldwide hydro-tectonic catastrophic condition and the consequent adjustments over the following thousands of years. Scientists interpret data easily through today's available evidence in geology, paleontology, ice age, climatology, atmospheric and numerous other scientific fields of investigation.
3   Provide a systematic historical timeline with archeological references and lawful infrastructural operational systems with scientifically identifiable evidence and socially reconstructable cause-effect relationships. Provide alternatives, social events from the beginning to our present time, natural laws, which lay at the foundation of the original creation.

The 'Young Earth Creation' scientists do not use geometric natural law or most of the 3DMM, yet the millennial Christendom did view the Christian Biblical creation week (Genesis 1:1 – 2:3) within the scope of geometric natural law and the infrastructural geometric solids. It becomes clear that this Creation week is not a random listing of mythological events. There is a clear engineering project management plan, schedule, activities, resource loading, procedures, quality assurance and control confirmations 'it was good', evidence of several foundational infrastructures upon which additional infrastructures have been erected. Figure 22 portrays an example of energy engineering; yet the same Figure and configurations apply to the complexity of creation – not only at the genetic and universal level, but also at the creative management levels (Figure 26).

In a similar way, Engineers and Logistics experts have conducted studies in other Biblical areas such as in the plans and dimensions of Noah's Ark. This engineering feat

Figure 26 - Genesis Creation: 7-day's Infrastructural GNL Approach

DAY	CORRESPONDING GEOMETRIC SOLIDS	EVENTS
1	**Tetrahedron** – three triangles	A) God creates Time – within infinity God creates sub-infinity, i.e., time + Geometric Natural Law (GNL) together. B) Heavens (dual) – quantification & extension – 3D (see 'deep' below next column) – the dual form of the term heaven suggests Creation of numeric concepts, 2 extensions -physical space (two areas and the heavenly above (angels – see Job 38:7). C) Firmness 'earth' – crossing of geometric fields – without form (waste), empty (no substance), waters (numerous quantities randomness); darkness; deep (3D) – spherical D) Spirit of God hovered over 'waters.' Light (energized darkness (structured emptiness). These are also designations for day and night – introduced movement in the randomness (waters) gravitational attraction (geometric – triangle): Energized – light; Division or distinction between empty space (darkness) and energized gravitated moving fields.
2	**Cube**, i.e., 4 sides; 1 top & 1 bottom squares	Expanse in the midst of the two areas of waters – above the expanse 'heaven' and that below the expanse: Waters being the myriads of crossing geometric fields are separated into two 'waters' – (geometric crossing fields) one above the expanse – called 'heaven,' and the other 'waters' below the expanse.
3	**Octahedron** (4 top & 4 BOTTOM triangles connected by a square)	Part 1 of 2. Genesis passage describes – waters under heaven to be collected in one place (dry land): Earth & collection of waters: Seas; Earth is to yield tender grass, herb-sowing seed, and fruit trees – after its kind. Geometric interpretation – here we have a crisscross of energized fields-pre-atomic, now within GNL form atomically organized to form matter, soil, dry land and seas. This energized land now has structure, elements to produce distinctive species (kinds) of plant life, herbs with its seeds, and fruit-trees. Here, life has its own genetic code, information, system, nano robotic engineering, conversion and reproductive systems.
4	**Octahedron**: 4 TOP & 4 bottom triangles connected by a square)	Part 2 of 2. In the heavens the 'luminaries' form a purposeful calendar for days, seasons and years. The great (Sun) light and the small (Moon) light to illuminate the Earth. Geometric interpretation: Levels luminaries purposes: distinguish between day/night, seasons (quarterly measure of Earth's orbit around the Sun), and year (God's calendar was focused on a solar year); no calendar measurement for the moon. However, the Moon is used as a night luminary along with the stars, while the Sun for daytime. These luminaries rule (manage by laws) in the sky.
5	**Icosahedron** 20 triangles and two pentagons arranged	Waters are to teem with living creatures, the Earth's heaven with flying creatures. Created are also monsters & creeping living creatures. All are to multiply and fill oceans and the

	in: 5 lower triangles connected via common pentagon to 10 middle triangles, which are in turn connected via common second pentagon to 5 upper triangles.	Earth's heavens. Geometric interpretation: The waters are filled with living creatures, great monsters and things that creep, while the air is filled with flying 'fowl' creatures. All are to multiply and fill the seas and the air.
6	**Dodecahedron** (Pentagons: 1 bottom, 5 middle bottom; 5 middle top and 1 top = 12)	Part 1 of 2, the earth brings forth living creatures each according to its species, cattle, creeping things, and beast of the earth – all within their species. Make man in God's image to rule over the creatures in the sea, air, earth. Geometric interpretation: These are 'living creatures' as distinct from the material creations. The living creatures have awareness, volition, emotions – they have the capacity to react as well as to various degrees, make 'decisions' among choices/options. Each creature is made according to its genetic code, species for specific purposes within the environment. Also on this sixth day, human beings have been endowed with management and creative responsibilities – just as God ('in God's image') to manage the planet – flora and fauna. For food, humans would receive nourishment from the fruit of every seed bearing herb and of the fruit-bearing tree, while living creatures from the seed bearing green herb plants.
7	**Dodecahedron** (Pentagons: 1 bottom, 5 middle bottom; 5 middle top and 1 top = 12)	Part 2 of 2, genesis description: God completed all His work by the seventh day. He blessed, sanctified the seventh day and rested. Geometric interpretation: The example here is that just as God worked during six days, He also fulfilled the blessed and sanctified day of rest. This is the relationship 6+1. The Hybrid Uniformitarians provide a paradox when they must somehow account for 'billions of years' of God's rest period. While uniformitarians in general should realize that not only do the six-day activities do not correspond with the Big Bang events, but in most cases the six-day activities are in reverse to those synthesized by the uniformitarians (see Figure 3)

provides key features in the reconstruction of Noah's Ark. Builders of Noah's Ark had to consider environmental challenges to overcome: fauna logistics, unprecedented radically changing global conditions, globally escalated tidal waves, temperature inversions, worldwide hydro-tectonic catastrophic events during an initial 40 days and lasting an additional 11 months of 'cooling' off. This included contrastive information on pre-, during and post Flood conditions. This comprehensive picture tracks data history and corroborating scientific geologic and paleontological evidence, an ice age when parts of the Earth's oceans froze (Job 38) (2000 B.C.). Worldwide climate adjustments after the ice age resulted in droughts that covered large sections of continents – areas that at those latitudes originally had abundant rains and wildlife

(Jacob's time – Genesis 41ff) (1,500 BC). This brought the beginning of large-scale ice-pack meltdowns that raised ocean water levels between 100 feet (north and south) and 500 feet (tropical zones).

After the historical Biblical events, Christians continued to maintain, expand and apply the Biblical knowledgebase (3DMM). Persecuted before Constantine and given Imperial power during and after Constantine, Christians had the opportunity to re-create a world Empire that removed pagan phallic worshiping anchors, addressed issues of pluralism, law, history, government, cultures, justice, commerce, banking, etc. Christianity, at that time, had substance and was set on establishing and developing an alternate world in contrast to the fallen pagan world. What is the substance that existed before and is now missing from contemporary Theological Curricula today? Today's Christianity must compete with Modernism, competitive religious pluralism, and after WWI and WWII has become impotent as moral authority.

How did the early Christians function in a secularize environment? In Western Europe St Ambrose and St. Augustine in the 5th century, the Celtic Orthodox in the West, and in Eastern Europe, Cyril and Methodius in the 9th century – this was a time when there was unity within Christendom. Christian missionaries dispersed throughout Europe and interfaced not only with their Church and monasteries but also with the city councils, kings, and people. Here they organized quality living on two fronts: a) the original Christian standard – spiritual, ethics (management style -executive), education (supervisory), behavior (functional); and b) implemented a Christian view in business practices, legal/court, non-usurious money management, logistics (movement of goods & services) to facilitate and maximize the contributions of the communities.

A debate about the relationship between Holy Traditions and/or Holy Scriptures can be viewed in terms of the establishment of a Knowledgebase - 3DMM, DMPS and qualitative vertical change (self-improvement and self-perfection). This is an area of the Management Style plans. Briefly - Tradition has a historical value, while the Scriptures are part of that long tradition. The knowledgebase and history had been maintained through Church chants and art - architecture, design, depiction of scriptures in 2D (icons, frescos) or 3D (statues, replicas), furniture and liturgical 'principles of drama' that function as memory activators of Christian traditions' world. We have a culture that reflects the works of the Holy Spirit (executive) (see Table 3). Above all, the designations of the scriptures as 'Old' and 'New' Testaments reflect the legal, inheritance transmission and management nature of this documentation (knowledgebase). These are covenant designations, and even more specifically – Marriage Covenants between the Eternal Lord God/Jesus Christ and the righteous Bride of Christ. Therefore, when Christians talk about the Holy Tradition and the Scriptures they are actually placing themselves – individually, as brothers, sisters, mothers, and fathers - the body of Christ, in a Marital Covenant. This Covenant is functional and has a specific purpose in the World. As a body of Christ, these traditions and the scriptures refer to precedents that have been provided by the Saints, and now its members as 'Saints-enhancers' towards each other – i.e., Christians work together to ensure that each one of them makes it to Heaven. In this Covenantal Marriage Christians must talk with the Christian's Bridegroom daily (prayer) and support and encourage Christian families and members to do the same (directly, in-directly). All this is part of the true Christian Tradition, and Christians recognize this in the bulk of the Scriptures. Both Tradition and the Scriptures become the Christian's very body, clothes and atmosphere of their daily living – this is how Christians become the salt and the light to the world.

Tradition is a social value. Tradition is the knowledgebase from which Christians derive their intellectual, physical, spiritual and cultural nourishment. This knowledgebase contains not only the best practices, examples, behavior, communication among members of the body of Christ but also the standard values and decision-making exercises. This includes conditions and precedents that lead to the various decisions that members and groups are likely to take. For example, what are the causes for delinquency, atheism, pre-marital activities, divorce, and loss of faith – there are key precedents in the Christian Traditions that show how 'Romans 1' activities and worldviews originate, are conceived, and entered into and the consequences expected? The body of Christ ensures to flag these precedents when taking a certain route. All this is known as wisdom. The Proverbs and Psalms contain a body of wisdom that King David applied constantly. He found nourishment from this 'wisdom' and it made him become a better king.

What is wisdom? It is the exercise of executive functions. As depicted in Table 3 in the 10 Commandments - the Beatitudes that are the core of the Traditions (wisdom). How diligent are Christians in maintaining this tradition / knowledgebase / wisdom. The more Christians remove themselves from this tradition the more they enter Romans 1 and Modernism, and the more they lose the urgent and critical need for repentance. Christians are their brothers/sisters' keepers.

This is what contributes to establishing a favorable culture where 3DMM, DMPS, OUSM thrive. Christians are not 'rugged' individualists who fight and take advantage of worldly opportunities. Christians must develop their professional and spiritual gifts to the highest level. They were to not only become useful in the world, but also invest their contribution into the Body of Christ. This is part of establishing the Management Operations and Style levels upon a firm Management Design. This then improves, enhances, and increases the quality of the Traditions for the future (within the scope of wisdom/holy spirit). Christian existence is to give, sacrifice and above all love what Christians have for Christ and He has for Christians. The choices Christians make, the values they nurture, their highest priorities must be in total conformity with the original traditions Christians have inherited and that which they will pass on. The language Christians use; the attitude and behavior they have is their external sign of our

Traditions and Scriptures they embrace daily. The True Christian Church, based on Holy Traditions and the Holy Scriptures, must become the most wonderful place to be on Earth! Similarly, as mentioned above – their contribution in the scientific, technological, bio, social and political sciences are based on Geometric Natural Law – evident in all creation: living world, family, groups, organizations, social, national and world levels. The Byzantine Empire maintained Greek discoveries and expanded them to the rest of the World, and became available to the Islamic world. In Europe, the Renaissance helped launch qualitative development in all branches of science, arts and politics.

How can we be sure that evolution is 'an almost certainty and true' when school textbooks attempt to 'prove' evolution through outdated, disproved and fraudulent 'scientific' ideas? Textbooks contain uniformitarian approved science such as peppered moths, chemical evolution, embryonic recapitulation, the horse series, and half a dozen other examples). Yet 20 years ago, the more honest evolution scientists have disproved these 'proofs.' Daily, the ideological uniformitarians scientists, media, curriculum designers and judges armed with a uniformitarian axiomatic posture use the UR, ER, SR, triple summersault logic, a reductionist 3-to-7 step scientific method (that ancient pyramid builders surpassed); and artificial or synthetic views of a simple-to-complex history. This results in two generations of public school and college graduates who think in terms of synthetic, non-existent realities and foundations. With this, they have learned how to daily brainwash themselves with cynical reduced simple-to-complex thought processes. They use syllogistic processes in an attempt to dismantle the 3DMM and misapply the DMPS.

Today, we have a new self-appointed class of Modernist high priests of Scientificism who function like amnesiacs because they have a synthetic rather than actual knowledge of history. Scientificists who have not been taught to reason from a third point of view (e.g., dialog) and methodologically are satisfied with 'almost certainty', not realizing that this is really uncertainty. How would such high priests who promote the uniformitarian gospel be able to distinguish between sanity and insanity when they hear ER edicts such as the notions that the 'theory of evolution has reached a point of almost being certainty and is truth?' It is like saying that a woman is almost pregnant, but everyone will consider her pregnant. When it is implied and the courts decree that 'science' is synonymous with 'Evolutionist Realism'(ER) – along the same line as that of Soviets claimed 'Socialist Realism.' That 'uncertainty' is inherent in hypothetical and theoretical conditions. Yet, being uncertain, the hypothesis and theory are not subject to DMPS stages. In this situation uniformitarian 'uncertainty' become almost certainty and truth by simply sitting around long enough. Then we have the situation where if anyone should disagrees or opposes this new UR, ER, SR 'scientific revelation' he/she/it would hereby be considered to be irrational, divisive and functions like a social 'vestigial organ.' Courts decrees enforce such uniformitarian truisms and force societies and organizations to reflect the UR/ER reality. All users of the UR/ER reality must filter this reality through the exclusive UR/ER filters. This is why its UR/ER users appear to talk in terms of slogans, canned logic, axiomatic, become confrontational, accusatory, patronizing and supremacist – until they run out of their repertoire of pseudo-scientific antics. Scientificists that do not compliance with UR/ER standard policy, procedures and rules are corrected or rejected from the UR/ER/SR inner sanctum. This is why we have social re-engineering, the criminalization of Christian ethics, practices and worldview. This is why the UR/ER seeks to re-drafting the American Constitution – not through majority rule but through activist court decisions.

Upon re-examining the UR/ER abysmal results on the *Scientific Report Card*, it is amazing how the Scientificist Pontificates manage. Obsessive naïve realist, their imaginary billion-year synthetic and artificial pre-history erected upon uniformitarian axioms and not upon scientific evidence. All this becomes clear when we examine Table 7 – *'Comparison: 4-Point Rule – Uniformitarian and Flat Earth views'* and find that there are precedents to this type of reasoning and the resultant failed civilizations. When objective researchers examine 'ER's low score and the multimillion genocidal reputations during the 20th century, most people will want to totally disassociate themselves from any ERs, since they may foresee potential International Tribunal action. It is clear that uniformitarian 'Evolutionist Realism' substantiated the foundation for World War I and II and continent-wide destructive revolutions –e.g., Marxists (division of classes), Nazis (division of races), Communist China (Cultural Revolutions) – whose political parties established their existence, policies and action upon UR/ER. In less than 50 years during the first half of the 20th century, this UR/ER platform sponsored at least half a trillion (500 million) deaths in Europe alone. Other multiple millions in the USSR and China died in uniformitarian court approved concentration camps, population movements and numbers' adjustments. Today, ER stands in the forefront as the legitimizing foundation for additional millions of human lives in areas of abortion (40 million in the USA), contraception, and sterilization (e.g., Mexico), as well as, in the experimentation with human embryonic stem cells. This research is done within the parameters of two ER scientific 'uncertainties' - when does human life begins, and its limited mental model which do not recognize information and nanobot functions at sub-chromosome levels. We have yet to see the effects of the depopulation schemes approved by the UN sponsored Club of Rome in 1975.

Now with such an abysmal score on the Scientific Score Card, the UR/ER members are appointing themselves and judges who will define what constitute science vs. pseudo science; science vs. religions and reality itself. This package, along with 'hate crime legislation,' redefinition of marriage and anti-Christian policy are financed and being exported under the umbrella of American Democracy that is also being exported as an Americanist resource to other countries around the world.

Upon yet a closer examination into history, it becomes evident that the 'Uniformitarian/Evolutionist Realist' doctrine is designed to throw societies back to the pre-scientific dark ages. This is a period in history where those who were in power had ensured their future by keeping the economically controlled and disinherited masses at a hand-to-mouth bestial existence subordinate to State's whims (see the five levels of econo-myth vehicles describes above (section 4.2 and Tables 7 and 9).

Similarly, millennial original Christianity went beyond the mere 'personal salvation' or 'give your heart to the Lord' convictions. Original Christianity contained within its legal historical and 3DMM and DMPS structures the elements of social and cultural rebirth or Renaissance - recognizing that man is more than an advanced animal. Man, instead has executive (spirit) and supervisory (soul) management infrastructures that can manage and create perfection within geometric natural laws. Mankind is made in the image of the supra-natural Eternal Lord God, and has the mission to contribute into the establishment of the Kingdom of God. This is mankind's mission to daily qualify as a citizen of the Kingdom of God. Thus the Christian Church had been moving humanity and civilization upward to that end.

The originators of the American Constitution had recognized these Christian Renaissance geometric natural law qualities, and implemented them in the new Republic's framework. Yet it is these qualitative elements that are being criminalized by the new 'democratized' American and European pagan and secular societies. George Orwell clearly warned of the detailed methods by which social re-engineering was to occur see his 'Animal Farm' and '1984.' These re-engineered doctrines and laws are being promoted as 'Freedom' from geometric natural law-based views.

Now under the redefined umbrella of separation of Church and State[lxxx] this allows for internal social and cultural changes (revolutions). The legal implementation of UR/ER/SR follows a well-beaten path. Begun with the outlawing of prayer in school and then any public institution, removal of the 10 Commandments, Christian symbols from public places, there appeared Malthusian depopulation schemes - in the USA - in 1970's two children per family limits, contraception, opening the doors to sexual promiscuity, easy divorce. Then with the passage of 'hate crime' legislation, which demonstrated the methodology for conducting unpopular legislation through courts and government institutions, this opened the door to wider road to the redefinition of marriage, implementation of abortion, fetus/embryonic stem cell research; criminalization of Christianity, court decisions on defining what constitutes science (UR/ER) vs. cultism.

What is curious about this State/court anti-Christian orientation is the emergence of secularist (pagan) policies, with its sidekick - the re-emerged religion known as 'phallic worship.' A closer examination reveals the numerous Executive Orders that have now been signed for implementation: eliminate private property, freeze human rights, etc.

CHAPTER 6 - WHAT IS THE BEST ANSWER?

6.1 – Two Alternatives for Action

If you agree or promote UR/ER policy, then just sit back, relax, play truth or consequences and follow events. You will, do doubt, track and/or maintain the debaters' development, publications and trends. On the other hand, if you see scientific and technical heritage slipping due to results on the *Scientific Score Card*, you may wonder if there is enough turn-around time left to correct the degrading situation in the scientific arena. Perhaps, you may seek out those who have not 'drunk from the cup of perdition' – and those who can still help revert to the objective geometric natural law-based reality.

6.2 – Initial 10-Point Action Plan

The uniformitarians are unable to provide scientific proof for their concept of evolution. For example, the book published in 1959 by Cardinal Ernesto Ruffini, '*The Theory of Evolution Judged by Reason*,' Joseph F. Wane, Inc; New York, NY, clearly shows that after 50 years of juggling and tossing the uniformitarians still fail to demonstrate a scientific evidence for evolution. At the same time, the same book has shown that over the same 50-year period, evidence for the recent Christian Biblical Creation model has stabilized. ICR's *Creation Scientific Model* addressed all dating methods and identified others, geologic morphology and plate tectonics, fossil interpretation, speed of light and gravitation, mutational effects, etc. have provided better scientific explanations and predictions. At the same time, some neo-evolutionists sought non-uniformitarian/ evolutionary explanations after they observed information processes and nano-robotic technology at sub-chromosome levels.

Yet, the above-mentioned second group will find an almost insurmountable task to correct this permeating two-century reductionist initiative. Those who wish to return to the constitution-based America, which provided the favorable conditions for the development of science - established upon its geometric natural law, the Renaissance, and the original and Christian Apostolic principles - must have a plan for re-designing the reality in its original model.

Those who allow for the pursuit of accurate knowledge, the scientific method must clearly:

1) Define what constitutes the 'objective or unfettered scientific method' (OUSM) in objective terms, such as those identified in the engineering project management approach.

2) Identify and formulate the current scientific methodology, its strengths and limitations, and clearly establish where this actually leads. Recognize the difference between the subjective, relative, reductionist approach that provides an ideological interpretation and method (e.g., dialectical materialism), in contrast to that incorporates the 3DMM, DBSP and other features described in this book. To recognize that the first has yet to yield concrete evidence for qualitative (vertical) progression; while the second predicts qualitative infrastructural values at all levels of science and seeks description, operations and validations.

3) Identify favorable conditions, where the scientific initiative thrives, and the legal historical geometric natural law-based methods such as those worked on by J. Kepler and Riemann. Recognize that today's Christianities are but shadows of what they originally had been during the past 2000 years -specifically, the favorable environments where we can recognize the pursuit for accurate science (3DMM, DMPS) and geometric natural law. We must re-assess historical data and information - much of which has been re-written.

4) Re-assess the true nature and direction of the scientific effort – its infrastructure, education, high-tech-based investment and scheduled discoveries.

5) Recognize and assess the low, medium and dramatic impact that certain doctrines have had upon the scientific, technological, educational, political, legal (philosophy and legislation), social, and faiths. Recognize the value of the shorthand formula:

$$\frac{\text{Ideology, Religion, Cults}}{\text{Geometric Natural Law, 3DMM Knowledgebase, Standards, Filters, DMPS, OUSM}}$$

6) Assess and correct the damage incurred to the scientific community, careers paths and opportunities of millions. Re-assess initiatives in the education system curriculum and policies; as well as, the political, economic, social, international, historical, standards and judicial systems, that have succumbed to ideological, legal manipulative methods, disinformation that had been applied under the various doctrines, e.g., UR/ER and 'Christian' pantheism and cultism.

7) Objectively reassess the qualifications of scientists, their authorities; institutions' accreditation criteria and value in all areas of the Evolution-Creation-ID-Hybrid Uniformitarian arenas of the debate. The initial legally-oriented questions for these answers should help uncover a Pandora's Box that up to now was kept closed by: media, money and resources, politics and policies, courts – e.g., the precedents of seemingly insignificant mini case-histories upon which the contemporary judgments are decided; the large-scale 'marketing' campaigns of disinformation; behind the scenes special interests and lobby activities.

8) Recognize that in many cases the decision-making and problem solving process has been short-circuited in many areas. Here we have the re-/definition of terms, issues filtered through questionable non-scientific filters. Qualified and objective scientists and administrators must take a closer look at the Gap Depth Determination to identify where gaps appear. Figure 18 will help identify where the contemporary scientific method fails. Uniformitarian court procedures and decisions have polarized instead of resolved issues. Ideologically rigged courts will not solve but further polarize the positions -courts are notorious for not being the bedrock of 'scientific' truth. Find objective criteria and documented evidence for qualified court determinations? If courts were established upon the original American Constitutional Law then justice, true science and religion/ideology based upon geometric natural law would reign supreme in righteousness.

9) Re-examine to what extent individuals, group and all, have been contaminated by 'ER' - in their priorities, behavior, budget allocation, scientists' qualification and promotions. Re-examine to what extent a variation exists in the use of Executive, Supervisory and Functional faculties (see Figure 5) – DMPS exercise. Resize, re-purpose, cleanup, reassessment, and establishment of accuracy-based knowledge, purposes, objectives and strategies. What models of thinking are we using? How do we view and interpret history and the knowledgebase?

10) Today, uniformitarian uncertainty causes never-ending debates in a pantheon of models: pantheism, theism, and supra-theism. One should ask who thrives in such an environment. It is clear that highly bankrolled relativists and 'Modernists' have much to gain in their algebraic world-view and a depopulated dark age.

George Washington's Vision describes the hope.

Appendix 1 - The 22 Journalistic Assumptions and Inferences

Some Journalists' questions are 'loaded' questions. These are questions that may imply a conclusion and really require to answer multiple implied questions (e.g., 22) yet this must be answered within a limited period (e.g., 30 seconds), or within one paragraph. In this case, it took a whole chapter 4 to address such a question, because it implied 22 inferences. This would be a typical an example of the violation of dignity, integrity and self-determination (S8).

In this instance, the question is '*Do you think that teaching evolution goes against religious beliefs?*' Here are the 22 inferences:

1.   A clear contrast exists between Evolution and Religion – issue of the 'Demarcation line'
2.   Evolution is not a religion
3.   Religion is not science
4.   Evolution and Religion are antonyms in their basic concepts
5.   Evolution is not a religion and not an ideology
6.   Evolution is scientific
7.   Religion is mythological and subjective
8.   Religion may or may not include Evolution
9.   Religion is not linked to reality as Evolution is (origins, supernaturalism vs materialism)
10. Religion is potentially anti-evolutionary and anti-science
11. Evolution is realistic, science, and rational
12. True science cannot exist outside the Evolutionary content or framework
13. Religion is abstract, unscientific and potentially irrational
14. Evolution is potentially, if not actually, anti-religious
15. Evolution may be included in religion, because of (#5), but Evolution is not a religion itself
16. Evolution describes a new reality, whereas religion has deep historical and cultural roots in mythology, folklore, work of fancy, or is a substitute for down to earth reality in the absence of evolution
17. Evolution is a recent discovery and has never existed in history in an ideological or religious form
18. Religion, in order to acquire a realistic foundation, must fully embrace Evolution.
19. The purposes, objectives and strategies of Evolution and Religion are dramatically different
20. The purposes, objectives and strategies of science and religion are dramatically different
21. The policies, procedures and rules of Evolution and Religion are different
22. To use any hypothesis with non-materialistic or reductionist perspectives is to be seen as being 'religious.'

Appendix 2 – Challenging Geological and Paleontological Evidence.

A perfect human footprint, a 14.5 inch long track, found embedded in a rock that an expert calculated from its location in the river's edge to be at least 150 million years Uniformitarian Time (UT) - 'The American Anthropologist' Volume IX, 1896, pg 68.

Dozens of other foot and shoe imprints have appeared up to that time, for example in the Fisher Canyon, Pershing County, Nevada, on January 25, 1927 – discovered a shoe print in an excellent state of preservation, heel edges smooth, rounded off as if cut, right side appeared more worn than the left (right foot). The rock, in which the heel mark was made, was Triassic limestone - 225 million years old (UT). Furthermore, micro-photographs revealed that the leather had been stitched by a double row of stitches, with double threads – a thread much smaller and refined than workmanship used by shoe-makers in 1927.

E.A. Allen, 'American Antiquarian', volume 7, 1885, page 39, and further detailed studies conducted in 1930 by Dr. Wilbur Greely Burroughs, head of the geology department at Berea College, identified twelve 9.5-inch man tracks among several creature tracks. The human prints were described as 'good-sized, toes well spread, and very distinctly marked' - located 16 miles east of Berea on Big Hill in Rock Castle County, Cumberland Plateau, impressed upon the gray Pottsville sandstone dating from the Upper Pennsylvanian period – over 300 million years (UT). Dr. Burroughs was quoted in the Louisville Courier-Journal, May 24, 1953: 'Of these, two pairs show the left foot advanced relative to the right. The position of the feet is the same as that of a person. The distance from heel to heel is 18 inches. One pair shows the feet parallel to each other, the distance between the feet being the same as that of a normal human being.' Dr. Burroughs concluded that the prints were made by a creature that was exclusively bipedal not quadruped and without evidence of tail acting as a tripod ('third leg') or signs of belly or tail marks in the examined stratum. Microscopic analysis of the human tracks, based upon grain count - 'the sand grains within each track are closer together than grains immediately outside the tracks and elsewhere on the rock for the same kind and same combination of grains, due to the pressure of the creature's foot.' The 'creature' exerted a weight pressure a little above that of a modern man. The clear impressions showed five toes, ball and heel totally unrelated to an amphibian's or reptile's physical makeup – only man has a foot like that. Albert G. Ingalls, 'Scientific American January, 1940, stated 'If man existed as far back as in the Carboniferous Period in any shape, then the whole science of geology is so completely wrong that all geologists should resign their jobs and take up truck driving.' On July 20 1968, the Cambrian shale (600 million years UT) revealed a five-toe child's 6-inch foot wearing a moccasin with trilobites for company - Dr. Clifford Burdick in Antelope Springs, Utah - Swasey Mountains and the Cambrian Wheeler shale. Similar discoveries made in areas of metallic objects with properties that confound UT timetables. For example, an iron 'cube' (2.64 x.2.64 x 1.85 inches, weight 1.73 lbs, specific gravity 7.75) was discovered at an Upper Austrian foundry in the fall of 1885, but in 1966-67, the iron 'cube' was subjected to electron-beam microanalysis by experts at the Vienna Naturhistorisehes Museum. The results: no traces of nickel, chromium or cobalt in the iron = not of meteoric origin. No sulfur detected = not a pyrite, a natural mineral that sometimes forms geometric shapes; low magnesium content. Dr. Kurat of the Museum, and Dr. R. Gill of the Geologisehe Bundesanstalt of Vienna, concluded that the object was made of cast-iron. In 1973, Hubert Mattlianer concluded from yet another detailed investigation that the object had been made from a hand-sculptured lump of wax or clay pressed into a sand base, forming the mold into which the iron had been poured. Conclusion: fabricated, but was encased in coal dating to the Tertiary - 60 million years old (UT).

INDEX

END NOTES

[i] The *Oronteus Finneus Map*, 1513 A.D and the *'Piri Re'is Maps'* depict detailed coast lines of every continent on Earth in longitudes and latitudes at a time when ocean levels were 100 to 500 ft lower than they are today. Researchers discovered these maps at the Turkish library – maps that depict detailed contours of world continents including that of the Antarctica continent before the glaciations of this continent. NASA's satellite radar scanning of Antarctica confirmed what this map depicted - rivers, mountain ranges, bays, etc. The *'Turk Ibn Ben Zara's'* (1487 A.D) maps depict maps of a post-Ice Age Mediterranean coastline. There are also other corroborating maps from *Oronteus Finaeus, Buache* and *Mercator Maps* (1700 A.D.) One can trace the source of these maps to the Phoenician mariners – who originally used and maintained these maps. This information about maps have been researched, among others, by Hapgood, Charles H. Hapgood, 'Maps of the Ancient Sea Kings' Kempton, Illinois, Adventures unlimited Press, 1996 ed., also by Nienhuis, James, I., 'Ice Age Civilizations,' Genesis Veracity, Houston, TX, 2006.

[ii] http://www.youtube.com/watch?v=8Uu-iPHJejs&feature=related

[iii] A.G. Drachmann, 'Heron's Windmill', *Centaurus*, 7 (1961), pp. 145-151

[iv] Dietrich Lohrmann, 'Von der ostlichen zur westlichen Windmuhle', *Archiv fur Kulturgeschichte*, Vo. 77, Issue 1 (1995), pp. 1-20 (10f.)

[v] http://en.wikipedia.org/wiki/Hero_of_Alexandria

[vi] Johnson, Robert, Bowie, Jr. 'Noah in Ancient Greek Art,' Solving Light Books, Annapolis, Maryland, 2007

[vii] The intellectual climate that led to this point can be traced to the writings of naturalists or Natural Philosophers. Here, Charles Linnaeus (1707-1778) speculated in his 'Botanical Philosophy' (1751) about the 'fixity' or immutability of species - later in his 'Systema Naturae' (1767), he considered more fluid, less fixed, specie limits. G. L. le Clerc Count of Buffon (1707-1788) attacked the fixity of species. Then Erasmus Darwin (1731-1802) dabbled with 'filament' or 'fiber' (inervous system) as the basis of all living things. E. Geoffroy Saint Hilaire (1772-1844) had later influenced Lamarck and Darwin, provided the first serious scientific exposition on evolution by considering a 'proto-homological' view, i.e., a 'structural plan' that revealed analogous organs. W. Goethe (1749-1832) in 1790 speculated on the 'metamorphoses' of plants and then of animals and man. George Cuvier (1792-1832) was a creationist – an opponent of Saint-Hilaire's ideas. Cuvier proposed multiple devastations of the Earth, by pointing to, and interpreting the multiple sedimentary layers. The Noachian Flood was but the last of these catastrophes. Then there was G. Le Monet de Lamarck (1744-1829) who provided a philosophical synthesis – 'spontaneous generation', in his lectures 'Systeme des Animau sans Vertebres' (1801) and his 'Philosophi Zoologique' (1809).

[viii] See also: J.W.G. Johnson, 'Evolution?' Perpetual Eucharistic Adoration, Inc, CA 1986; and Gerard J. Keane, 'Creation: Rediscovered,' Tan Books and Publishers, Inc. IL, 1999.

[ix] Atiyah, Michael and Sutcliffe, Paul, 'Polyhedra in Physics, Chemistry and Geometry,' Milan JH. Math 71:33-58.

[x] G.B. Babrymple, 'The Age of the Earth' (1991), Stanford, CA, Stanford University Press), p. 91; and G.B. Barymple, '40Ar/36 Analyses of Historic Lava Flows,' Earth & Planetary Science Letters', 6 (1969); pp 47-55.; Snelling A.A. 'The Cause of Anomalous Potassium-Argon 'Ages: for Recent Andesite Flows at Mt. Ngauruhoe, New Zealand, and the Implications for Potassium-Argon 'Dating', ' R.E. Walsh, ed., Proceedings of the Fourth international Conference on Creationism (1998, Pittsburgh, PA, Creation Science Fellowship), pp. 503-525.; Snelling, A., A. 'Excess Argon': The 'Achillies' Heel' of Potassium-Argon and Argon-Argon 'Dating' of Volcanic Rocks, (http://www.icr.org/article/436)

[xi] Example: H. Arp, Quasars, Redshifts, and Controversies (Berkeley, CA: Interstellar Media. 1987), and Seeing Red: Redshifts, Cosmology, and Academic Science (Montreal, Canada: C. Roy

Keys, Inc., 2002). E. Lerner, 'The Big Bang Never Happened' (New York: Random House, 1991)

xii Macdougall, Doug, 'Frozen Earth: The Once and Future Story of Ice Ages,' University of California Press, 2004

xiii 'How are past temperatures determined from an ice core? ' Scientific American, September 20, 2004 – http://www.sciam.com/print_version.cfm?articleID=00001580-C282-1148-828283414B7F012B)

xiv Kennett, J.P. 1982, 'Marine Geology,' New Jersey, Prentice-Hall, p. 747

xvi Milankovitch, M, 1930. 'Mathematische klimalehre und astronomische theorie der Klimaschwankungen.' In: Handbuch der Klimatologie, I.W. Koppen and R. Geiger (Eds.) Gerbruder Bontraeger, Berlin. Berger, A., S.H. Schneider, and J.C. Duplessy, eds., 1989. 'Oceanic Response to Orbital Forcing in the Later Quaternary: Observational and Experimental Strategies,' in' Climate and the Geosciences, Kluwer Academic Publishers, Dordrecht.; see also problems with astronomical theory: Vardiman, L., 1993. 'Ice Cores and the Age of the Earth.' Institute for Creation Research Monograph, San Diego, CA.

xvii 'A New Theory of Glacial Cycles' http://muller.lbl.gov/pages/glacialmain.htm;

xviii Holton, J.R., 1972, An Introduction to Dynamic Meteorology,' New York, Academic Press, pp. 48-51; Oard, Michael, 'The Ice Age and the Genesis Flood,' http://www.icr.org/article/272

xix http://www.icr.org/article/272/, http://www.icr.org/article/355/ , http://www.icr.org/article/2836/ , http://www.icr.org/article/120 , http://www.icr.org/article/383 , http://www.icr.org/article/390 , http://www.icr.org/research/index/researchp_as_platetectonicsl/ , http://www.icr.org/research/index/researchp_lv_r02/

xx Alley, R.B., et al., 'Visual-Stratigraphic Dating of the GISP2 Ice Core: Basis, Reproducibility, and Application,' Journal of Geophysical Research' 102, C12 (1997), pp. 26, 367-26,381; also Meese, D. A. Gow, A.J., Alley. R.B., Zielinski, G.A., Grootes, P.M., Ram, M., Taylor, K.C. Mayewski, P.A., and Bolzan, J.F., 'The Greenland Ice Sheet Project 2 Depth-Age Scale: Methods and Results,' Journal of Geophysical Research' 102, C12 (1997), pp. 26,411-26, 423; also Oard, Michael, 'Are Polar Ice Sheets Only 4500 Years Old?' http://www.icr.org/article/120/

xxi Vardiman, L. 'Numerical Simulation of Precipitation induced by Hot Mid-Ocean Ridges,' http://www.icr.org/research/index/researchp_lv_r04/ ; Vardiman, Larry and Bousselot, Karen, 'Sensitivity Studies on Vapor Canopy Temperature Profiles,' http://www.icr.org/research/index/researchp_lv_r05/ , see also results: Wise, K.P., S.A. Austin, J.R. Baumgardner, D.R. Humphreys, A.A. Snelling, and L. Vardiman, 'Catastrophic Plate Tectonics: A Global Model of Earth History,' Proceedings of the Third International Conference on Creationism. R.E. Walsh, et al., Editors, 1994, Creation Science Fellowship, Inc., Pittsburgh, PA, pp. 609-622. Also: Braumgardner, John R. 'Runaway Subduction as the Driving Mechanism for the Genesis Flood,' http://www.icr.org/research/index/researchp_jb_renawaysubduction/.

xxii Refer back to Chapter 1 where the works of John William Draper with his book 'History of the Conflict between Religion and Science' (1874); and later Andrew Dickson White with his essay 'A History of the Warfare of Science with Theology in Christendom' have been brought; the Vienna Circle and Karl Popper and others have been brought.

xxiii Alan, B. Spitzer, 'Historical Truth and Lies about the Past.' University of North Caroline Press, Chapel Hill, 1996. Keith, Jenkins, ed. 'The Postmodern history Reader.' Routledge, London, 1997 xxiv See the many articles published on www.ICR.org, and others, such as: 'Journal of Creation' http://creationontheweb.com/content/view/3873/ (previously titled: 'Creation Ex-Nihilo Technical Journal')

xxv A similar process emerged at the end of the Moslem Classical period, where Al-Biruni who introduced the 'demarcation line' between theology and empirical science (Ahmad Dallal,' Encyclopedia of the Quran, 'Quran & Science'). Similarly, Ibn Tufail's philosophical novel, translated as 'Philosphus Autodidactus' (1671) introduced empiricism, tabula rasa, materialism (G.A.Russell (1994), The 'Arabick' Interest of the Natural Philosophers in Seventeenth-Century England, Brill Publishers; http://en.wikipedia.org/wiki/Islamic_Golden_Age

xxvi On a continuous basis, the media presents the public with paleontological or geological proof that should reinforce the long ages bias. While writing this book the following information was released via internet: 'Japanese and Mongolian scientists have successfully *recovered* the complete skeleton of a 70-million-year-old young dinosaur, a nature museum announced Thursday.' http://news.yahoo.com/s/ap/20080724/ap_on_re_as/japan_dinosaur

xxvii 'Expelled: No Intelligence Allowed' starring Ben Stein, 2008

xxviii See: Luisi, Pier L. (2006) 'Emergence of Life: From Chemical Origins to Synthetic Biology.' Cambridge University Press. http://en.wikipedia.org/wiki/Abiogenesis.

xxix http://www.icr.org/article/3932 ; this program is feely available for personal use and can be downloaded from the web at http://mendelsaccount.sourceforge.net; also, Snelling, Andres, A, ed 'Proceedings of The Sixth International Conference on Creationism: Technical Symposium Sessions,'Creation Science Fellowship, Inc Pittsburgh, PA and Institute for Creation Research, Dallas, TX., 2008.

xxx The reasoning follows that Natural Selection describes the condition of heritable traits through successful generation of population of reproducing organisms, acts on the phenotype (observable characteristics of an organism), which with genotypes associated with the favorable phenotypes increase in frequency and the following generations and adaptations will occur in an ecological niche, thus giving rise to a new species

xxxi Ibid. http://www.icr.org/article/3932

xxxii 'But I say to you that for every idle word men may speak, they will give account of it in the Day of Judgment.' (Mat 12:36); '...but man lives by every word that proceeds from the mouth of the LORD' (Deut 8:3).

xxxiii Voet, D. and Voet, J.G., 1995. 'Biochemistry,' John Wiley and Sons, Inc., New York, p. 21. Orgel, L.E., 1994. 'The origin of life on the Earth,' - Life in the Universe, Scientific American, 271 (4):53-61.

xxxiv Siddle, H. et. Al, 2007. 'Transmission of a fatal clonal tumor by biting occurs due to depleted MHC diversity in threatened-carnivorous marsupial.' Proceedings of the National Academy of Science,' 104(41):1622-16226.

xxxv Schmid, R. 'Cancer forces Tasmanian devils to breed earlier.' Associate Press, July 15, 20008 (web accessed July, 2008); also Jones, M., et al. 2008. 'Life-history change in disease-ravaged Tasmanian devil populations.' Proceedings of the National Academy of Science, July 14, 2008 (ahead of print. Accessed July 17, 2008)

xxxvi Frank Sherwin, MA, 'Tasmanian Devils: Extinction not macro-evolution,' http://www.icr.org/article/3870

xxxvii Frank Sherwin, ibid.

xxxviii An classic example among many: Henke, K.R., Young-Earth creationist Helium Diffusion 'dates' (March 17, 2005 at http://www.talkorigins.org/faqs/helium/zircons.html; Henke, K.R. 'young-earth creationist helium diffusion 'dates' (November 24, 2005) http://www.,talkorigins.org/faqs/helium/zirocons.html; and rebuttal by Dr. Russell Humphreys, http://www.talkorigins.org/faqs/helium/original.html; http://www.trueorigin.org/helium01.asp ; and http://www.trueorigin.org/helium02.asp .

xxxix Mark Isaak's 'Problem with a Global Flood,' Second Edition, The Talk Origins Archive, November 16, 1998 – http://www.talkorigins.org/faqs/faq-noahs-ark.html ; and the rebuttal by J. Sarfati,'Problems with a Global Flood?'1998, http://www.trueorigin.org/arkdefend.asp

xl 'Mendel's Accountant: A New Population Genetics Simulation Tool for Studying Mutation and Natural Selection' by Dr. Baumgardner; and 'Using Numerical Simulation to Test the Validity of Neo-Darwinian Theory' by Dr. John Sanford http://www.icr.org/article/3932 ; this program is feely available for personal use and can be downloaded from the web at http://mendelsaccount.sourceforge.net

xli http://www.icr.org/article/darwinian-medicine-prescription-for-failure (accessed March 01, 2009)

xlii For example, the Department of Justice may wish to check all correspondence that went on five years ago to determine whether Company management had been aware of the product's safety or health hazard that the company produced.

xliii NSA, 500 fifth St. N.W. Washington, D.C.
http://books.nap.edu/openbook/php?chapselect=yo&page=01&record_id=11876.

xliv http://www.trueorigins.org/arkdefen.asp

xlv http://www.wikipedia.org.

xlvi By the National Academy of Science (NAS) 500 fifth St. N.W. Washington, D.C. http://books.nap.edu/openbook/php?chapselect=yo&page=01&record_id=11876

xlvii Gregor Johann Mendel (1822 – 1884) an Augustinian priest and scientist is considered to be the father of modern genetics. Genetics is the study and discovery of consistencies, laws of inheritance of traits in peas and attempted in honeybees. He presented his paper at the Natural History Society of Brunn in Moravia (1985) and published this paper a year later in the Proceedings of the Natural History Society of Brunn.

xlviii The Associated Press story, on June 26, 2008, reported the finding in Latvia, of a supposedly 365 million year old fossilized four legged water creature – 'Ventastega' which is now identified as a 'dead-end' because several 'older' tetrapods with more advanced physical features have been found earlier. So 'Ventastega' could not qualify as a 'missing link' it was 'out of sequence in timeing' (Neil Shubin, professor of biology and anatomy at the University of Chicago). Such discoveries don't match with the uniformitarian geologic column, but does fit perfectly within the creation global catastrophe events.

xlix http://news.bbc.co.uk/2hi/uk_news/magazine/7540427.stm;
http://ncseweb.org/rncse/21/3-4/flat-earth-society-president-dies; Flat Earth FAQ, http://theflateearthsociety.org/forum/index.php?topic=69.0

l ICR scientists have successfully addressed all of these areas. Readers should be encouraged to visit ICR website at www.icr.org. It is curious that evolution scientists appear to be ignorant of this vast amount of researched literature that has provided alternate successful and better scientific explanations.

li Goldschmidt, R., 'The Material Basis of Evolution,' New Haven, Conn., Yale University Press, 1940; Gould, S. J., 'The Return of *Hopeful* Monsters,' Natural History, June/July, 1986 http://www.stephenjayghould.org/library/gould_hopeful-monsters.html

lii Niles Eldredge and Stephen Jay Gould, 'Punctuated equilibria: an alternative to phyletic gradualism' 1972, (http://www.blackwellpublishing,com/ridley/classictexts/eldredge.asp).

liii Ernst Mayr, 'Speciatonal Evolution or Punctuated Equilibria'
(http://www.stephenjaygould.or/library/mayr_punctuated.html).

liv 'A medical imaging technique that generates a three-dimensional view of some object by combining a series of two-dimensioned X-ray image of 'slices' of that object.' (from NSA booklet, page 38.)

lv See research published on www.icr.org

lvi See Chapter 5 on what constitutes 'theism' and religion. Also refer to Table 8 'Religion and Ideology: similarities.'

lvii Figure 11: Comparison: DMPS vs. Uniformitarian 'Scientific Method'

lviii The concept of Teleology precedes Darwinian evolution. 'Telos' (Greek) for 'end, purpose' implies that everything has a design and that this design reflects a purpose – a bird as an aircraft are designed to fly through the air – the design implies the purpose. In other words, there's an inherent final cause (purpose) – 'telos' in every design. Where Plato and Aristotle identified a final cause; Lucretius (Democritus) conceived of metaphysical naturalism (materialism) or accidentalism -- precursor of uniformitarianism. In the modern world, teleology can be seen as thermodynamics in physics; fitness of elements within a complexity framework in chemistry; vital force in biology; process-driven systems (teleonomy) in communication and control.

lix Refer to page 96 where 18 examples of false science are magnified during many debates.

lx Supreme Court rule in Epperson v. Arkansas case (393 U.S. 97 – 1968) invalidating laws that prohibited the teaching of evolution in the classroom – issue of Establishment Clause; Edwards v Aguillard (482 U.S. 578 – 1987) Louisiana – ruling that invalidated laws that required equal time for teaching creationism and evolution; McLean v Arkansas Board of Education, 529 F. Supp. 1255, US District Court, 1982 – Establishment Clause; Kitzmiller v. Dover Area School District (Case # 04cv2688) – Federal Court ruled that Intelligent Design is not appropriate for inclusion in science classrooms and is essentially religious in nature. Selman v. Cobb Country School District (2005, Georgia) – disclaimers stickers to be placed into biology books that stated that evolution was only a theory and not a fact, was ruled by Federal District Judge Clarence Cooper to be unconstitutional and in violation of the Establishment Clause of the US Constitution.

lxi See also Table 6 'Comparison: DMPS vs. Uniformitarian Methods'

lxii The www.dictionary.com is selected for the following reasons: 1) conforms with the lexical and source standards used by other authoritative dictionaries, encyclopedias and thesaurus; 2) provides a fresh approach to definitions; 3) terms are contextualized in a comparative framework, for example religion is not primarily established upon a belief in a supernatural entity – religions can also be 'atheistic' – for example, reflect a focus on behavioral ethics and rule-based structures (e.g., Buddhism). Comparisons can be made within synonymic derivations – e.g., religion and ideology.

lxiii The term 'evolution' has not been used but the processes have been described: Thales suggests that all things originated from water; Anaximenes refers to the thickening and thinning of the air principle; Anaximander focuses on the concept of development from moisture under the influence of warmth, suggests long periods or change between animal to man. Empedocles, Epicarus and Lucretius build on this and suggest a type of natural selection. Heraclitus conceives a teleological rational development involving air to water and to earth with the involvement of the Logos who refines the process. See also http://www.iep.utm.edu./e/evolutio.htm.

lxiv Ernest L. Abel, Ancient Views on the Origin of Life (Fairleigh Dickinson University Press, 1973); also Stanley L. Jaki, 'Science: Western or What?' Intercollegiate Review (Vol. 26, Fall 1990); also, James Lovelock, The Ages of Gaia (New York: W.W. Norton and Co., 1988); also, Wolfgang Smith, Teilhardism and the New Religion (Rockford, Illinois: Tan Books and Publishers, 1988).

lxv Robert Muller, as cited in 'United Nations' Robert Muller – a Vision of Global Spirituality,' by Kristin Murphy, The Movement Newspaper, September 1983, p. 10.

lxvi http://www.canto.ru/calendar/index_en.php is a calendar that contains the history and names of saints throughout Christian Orthodoxy including the Celtic world. This is one of the most complete and comprehensive records among many others

lxvii During the European Classical period – Aristotle's foundational tools (algebraic) and Plato (geometric).

lxviii NOTE: The mathematical discipline of algebra should not be confused with the ideological implications that are being ascribed to Algebra.

lxix NOTE: In other words, the true position of the algebraic 'point' is really contingent on at least five geometric conditions within the infinite circle:

1) Time (diagonal AB)

2) Extension (space) on either side of the temporal diagonal AB

3) Numerics, and

4) A second 'fold' creating diagonal CD, and 5) at the cross roads of AB and CD there is the 'point' E).

lxx Quickly reference chapter 2 where evolutionist realisms - orchestrated by journalist Henry Louis Mencken and the ACLU in 1925 during the 'Scopes Trials' compare with the dynamics of Soviet doctrine of 'Socialist Realism.'

lxxi As mentioned earlier, the works of Plato have described this non-Biblical approach to geometric natural law.

lxxii See Book of Exodus 25:8 ff, where there is a description of the Tabernacle's layout and features. The 3MM can be derived from this layout and objects, where all of these correspond to the Management Design of the 3DMM (see Figure 1). The Operational and Stylistic management levels are derived through the corresponding activities, performance, judgments, attitudes, outcomes and others that follow.

lxxiii See the Tabernacle built by Solomon (Book of 1st Kings 8:4 and following). This Temple contained the Ark of the Covenant but the ark itself only contained the 10 Commandments provided by Moses, and neither the rod of Aaron nor the dish of manna

lxxiv Briefly, these are summarized in the Book of Deuteronomy 26-28 – contains a listing of the blessings and curses for following or not following the Decalogue, among others

lxxv Note that Jesus Christ commissioned his apostles and disciples to go to the 'lost sheep of Israel' - (Matthew 10:5-6). He recognized that he has been sent to the same 'lost sheep of the house of Israel' (Matthew 15:24). Curiously, we can find that it is most of the European nations that have converted to Christianity and proclaimed Jesus Christ and His Kingdom mission to the world for 2000 years.

lxxvi Genesis 2:7

lxxvii Musgrave, I: Baldwin, R. et al (2005) 'Information Theory and Creationism,'
(http://www.talkorigins.org/faqs/information/infotheory.html) (Retrieved 2008-03-24.

lxxviii Bergstrom, CT; Lachmann, M (2006). 'The fitness value of information.'
(http://arxiv.org/pdf/q-bio.PE/0510007

lxxix http://www.gallup.com/poll/108226/Republicans-Democrats-Differ-Creationism.aspx

lxxx The 'separation of Church and Sate' clause is not in the original American Constitution. This original American Constitution reflects the dynamics of a Republic: based upon Renaissance, Apostolic Christian and geometric natural law structures. This is what the 'Society of Cincinnatus' had been leading. Now America has the UN Charter to contend with – founded upon International Socialist Democratic rules, algebraic, subjective structure (UR).

The American Constitution specifically identifies the 'God' – as being 'Nature's God' – i.e., the Creator of Nature, and the State cannot escape being in the Creator's nature.

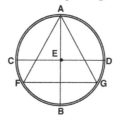